Thinking Like an Engineer

MW00490827

Thinking Like an Iceberg

Olivier Remaud

Translated by Stephen Muecke

polity

Originally published in French as *Penser comme un iceberg* © Actes Sud, 2020

This English edition © Polity Press, 2022

Excerpt from *Arctic Dreams* by Barry Lopez reprinted by permission of SLL/Sterling Lord Literistic, Inc. Copyright 1986 by Barry Lopez.

Polity Press
65 Bridge Street
Cambridge CB2 1UR, UK

Polity Press
101 Station Landing
Suite 300
Medford, MA 02155, USA

All rights reserved. Except for the quotation of short passages for the purpose of criticism and review, no part of this publication may be reproduced, stored in a retrieval system or transmitted, in any form or by any means, electronic, mechanical, photocopying, recording or otherwise, without the prior permission of the publisher.

ISBN-13: 978-1-5095-5146-0 (hardback)
ISBN-13: 978-1-5095-5147-7 (paperback)

A catalogue record for this book is available from the British Library.

Library of Congress Control Number: 2021949632

Typeset in 11/13pt Sabon
by Cheshire Typesetting Ltd, Cuddington, Cheshire
Printed and bound in Great Britain by CPI Group (UK) Ltd, Croydon

The publisher has used its best endeavours to ensure that the URLs for external websites referred to in this book are correct and active at the time of going to press. However, the publisher has no responsibility for the websites and can make no guarantee that a site will remain live or that the content is or will remain appropriate.

Every effort has been made to trace all copyright holders, but if any have been overlooked the publisher will be pleased to include any necessary credits in any subsequent reprint or edition.

For further information on Polity, visit our website:
politybooks.com

Contents

The frozen ocean itself still turns in its winter
sleep like a dragon.
— Barry Lopez, *Arctic Dreams*

Acknowledgements

I would first like to thank Stéphane Durand at my French publisher, Actes Sud, for welcoming this book into his 'Mondes sauvages' series and for following every step of the process with attention and friendship. I also thank Stephen Muecke for his translation into English and Elise Heslinga at Polity.

For their help in various ways (bibliography, translation, proofreading, illustrations, conversations), my gratitude goes to Glenn Albrecht, Þorvarður Árnason, Caroline Audibert, Petra Bachmaier, Chris Bowler, Aïté Bresson, Garry Clarke, Stephen Collins, Julie Cruikshank, Philippe Descola, Élisabeth Dutartre-Michaut, Katti Frederiksen, Sean Gallero, Samir Gandesha, Shari Fox Gearheard, Hrafnhildur Hannesdóttir, Lene Kielsen Holm, Cymene Howe, Nona Hurkmans, Guðrún Kristinsdóttir-Urfalino, José Manuel Lamarque, David Long, Robert Macfarlane, Andri Snær Magnason, Rémy Marion, Christian de Marliave, Markus Messling, Éric Rignot, Camille Seaman, Charles Stépanoff, Agnès Terrier, Torfi Tulinius, Philippe Urfalino, Daniel Weidner and Stefan Willer.

Finally, I am indebted to the Alexander von Humboldt Foundation, the Leibniz Zentrum für Literatur und Kulturforschung in Berlin and the Institute for the

Humanities at Simon Fraser University in Vancouver for allowing me to present parts of the manuscript as I was preparing it.

The Issue

Icebergs have been considered secondary characters for a long time now. They made the headlines when ships sank after hitting them. Then they disappeared into the fog and no one paid them any more attention.

In the pages that follow, they take centre stage. Their very substance breathes. They pitch and roll over themselves like whales. They house tiny life forms and take part in human affairs. Today, they are melting along with the glaciers and the sea ice.

Icebergs are central to both the little stories and the big issues.

This book invites you to discover worlds rich in secret affinities and inevitable paradoxes.

There are so many ways to see wildlife with new eyes.

The Issue

Prologue

They are Coming!

The morning was dark. Fog was suspended over our heads. Pancakes of ice floated near the ice edge. The sea seemed sluggish.

Then a discreet sun lit up the horizon.

Three points appeared in the distance. A thin silhouette emerged from the fog. I could not immediately identify the shape, but it was becoming more and more curved. No whale has these spurs on its back; my nomadic brothers are larger.

The clouds began to glow.

A ship was approaching us.

It was making slow progress. Like a lost penguin, it took small steps sideways. When it anchored in our vicinity, I saw them stirring. They were huddled together on the forecastle, jumping up and down in a strange dance. They were pointing at me. Their faces were long, their beards shaggy, and they smelt strong. They looked like ghosts. I could only make out the males. Some smiled, others opened their mouths but no words came out. With their hands on the main mast, some were kneeling and bowing their heads. They crossed themselves as they stood up.

A man emerged from a cabin at the back of the ship. He climbed the stairs leading to the deck. A group followed him. Drumbeats echoed in the silence of the

1

ocean. When the music stopped, he was announced by one of his companions.

Captain James Cook looked at the assembled crew and then addressed his sailors. His clear voice carried a long way. He told them that they had sailed far and wide, so far across the ocean at this latitude that they could no longer expect to see any more dry land, except near the pole, a place inaccessible by sea. They had reached their goal and would not advance an inch further south. They would turn back to the north. No regrets or sadness. He prided himself on having fulfilled his mission of completing his quest for an Antarctic continent. He seemed relieved.

As soon as the captain's speech was over, a midshipman rushed to the bow. He climbed over the halyards and managed to pull himself up onto the bowsprit. There, balancing himself, he twirled his hat and shouted, 'Ne plus ultra!' Cook called the young Vancouver back to order, urging him not to take pride in being the first to reach the end of the world. Screaming thus in Latin that they would go 'no further!' made him unsteady over the dark waters. He could fall into oblivion with the slightest gust of wind. The crew burst out laughing. With a smile on his face, the reckless hopeful returned to the bridge like a good boy. Then they turned their backs on me and went back to their tasks, some disappearing into the bowels of the ship while others climbed up to the sails.

Those three words echoed in the sky. I remember it with pride.

Call me 'The Impassable'.

I am the one who stopped Cook on his second voyage around the world, the happy surprise that cut short his labours at 71° 10' latitude south and 106° 54' longitude west.

I am one of the icebergs on which the Resolution, *a*

three-masted ship of four hundred and sixty-two tons, would have crashed if the fog had not cleared. On that day, 30 January 1774, they saw me in all my imposing, menacing volume.

My comrades from Greenland are slender. I am flat and massive. I blocked the way without giving them the chance of going around me. In any case, there is only ice behind me, an infinity in which they would have become lost. I saved them from a fatal destiny.

Thanks to me, an entire era thought that no one before the captain had gone so far south, that he was the sole person, the only one, the incredible one to have achieved this feat. What can I say about the snow petrels that have been landing on my ridges for centuries? I am familiar with these small white birds with black beaks and legs. They are attracted by the tiny algae that cling to my submerged sides.

Cook and his sailors kept their distance. Except for the times when they took picks and boarded fragments of iceberg from longboats. They climbed over them, dug them up and extracted blocks of ice which they left in the sun on the deck of the big ship to melt so they could drink their water.

We were much more than their tired eyes could count, not ninety-seven but thousands, an ice field as far as the eye could see.

We were a whole population.

1

Through the Looking Glass

A painter and a priest are standing at the rail of a steamer, the *Merlin*, on the way to the coast of the island of Newfoundland. They had left the port of Halifax, Nova Scotia, in the middle of June 1859 and are making their way to one of their destinations, Saint John. Having reached the foot of Cabot Tower, they meander north of the Avalon peninsula, between the Gulf of St Lawrence and Fogo Island, an area where strangely shaped blocks from Greenland are drifting. After about ten days, they embark on a chartered schooner called *Integrity* and sail towards the Labrador Sea. A rowboat is waiting on deck between the gangways that connect the forecastle and the stern. This will be their way to approach the giants.

Thus begins a chase that lasts several weeks.

A game of hide and seek

They are iceberg hunters.

They are armed with a battery of brushes and pens. Their satchels are overflowing with notebooks and drawing boards. Pairs of telescopic-handled theatre binoculars sit atop crates of paintings. Frederic Edwin Church intends to capture the volumes and colours of icebergs in oil studies and pencil sketches. He has a large work in mind. Louis Legrand Noble, on the other hand, is keeping a chronicle of their expedition. He wants to write a truthful account of it. The two friends play cards with other passengers. They reminisce, discuss the colour of the water and squint at the sky to judge the weather. They wait for the moment when they can see the faces of the 'islands of ice', as Captain Cook called them, up close. They are on the lookout, as eager as trappers, for an unusual catch. They are on guard, day and night, sleep poorly and flinch at the slightest sign. The swell makes their stomachs groan.

They made inquiries before leaving. They know that icebergs are a sailor's nightmare.

For the past ten years or so, the northern latitudes have been the focus of attention. HMS *Erebus* and HMS *Terror*, the two warships that Sir John Franklin commanded in 1845 in an attempt to open the Northwest Passage, have been lost. Jane Griffin, otherwise known as Lady Franklin, moves heaven and earth to find her husband. She convinces the British Admiralty to mount several search expeditions. Other governments quickly follow suit. The physician and explorer Elisha Kent Kane publishes two first-person accounts of the campaigns organised by the American businessman and philanthropist Henry Grinnell. His descriptions of desolate Arctic landscapes provided a stock of images that inspired an entire generation.[1]

Everyone wants to know what happened to Franklin. Significant economic and political interests come into play. Curiosity becomes bankable. Public opinion is intoxicated. A nation's reputation depends on this desire to know. But the research stalls. Until the mystery suddenly becomes clearer. In the spring of 1859, Francis Leopold McClintock, a regular member of the Royal Navy, and his officers collect evidence from an Inuit tribe on King William Island. They add more evidence and eventually find scraps of clothing, guns, bodies, a cairn, a small tent and a tin box on the ground with a clear message: the two ships had been icebound on 12 September 1846 and Franklin had given up the ghost on 11 June 1847. After this fatal winter, the survivors had decided, on 22 April 1848, to set out on a journey over the ice pack in an attempt to reach more hospitable lands. No one returned.[2]

Apart from a few minor scares, Church and Noble's journey goes off without a hitch. The skies are clear, the sea is friendly. One fine day, the deckhand calls out: 'Icebergs! Icebergs!' Relief and euphoria: their goal is in sight. The passengers move towards the bow. Two elegant masses of unequal size emerge. The ship is slowly approaching the bigger one. The companions' eyes widen. But a thick fog starts to spread. Clouds fall over the sea like a stage curtain. They cover the horizon and the show comes to an end. Having their final act taken away, the travellers are disappointed, almost offended.

During a stopover on land, fishermen explain to them that iceberg hunters must be patient. It is always a game of hide and seek. In this game, the roles are unequal and the rules are constantly changing. Icebergs know the winds and currents better than humans. They are mischievous and do not let themselves be caught. They disappear as suddenly as they reappear. If you get too close, they run away or get angry. They are more intelligent than their pursuers.

The icebergs have made a pact of friendship with the fog. No one can break it. When the clouds transpire, water droplets become ice crystals that pile up on top of each other. Then these crystals return to the clouds as they evaporate. In the meantime, the blocks take advantage of the moments when the air becomes thick with moisture to escape from view. Icebergs and mists unite the sky and the sea. Their relationship is mutual. Each partner benefits. By way of encouraging them to turn back, the fishermen tell our two dilettantes a secret worthy of the best pirate stories: 'No jackal is more loyal to its lion, no pilot fish to its shark, than the fog to its berg.'[3] A chill runs down Church and Noble's spines: they understand that, in such reciprocal living pairs, the iceberg is the predator. Mists follow it everywhere. They are inseparable.

At the beginning of July 1859, a group of thirteen icebergs encircles the schooner. The painter and the narrator are ecstatic. They will finally be able to examine them closely. The boat is lowered. With the necessary care. When icebergs roll over, they take everything with them in their chaotic movements and cause panic around them. Sections of ice can collapse and crush the boat. The captain on board orders the rowers to keep a respectable distance.

For several minutes they make their way through the floating masses, taking advantage of a clearing in the sky and a calm sea. They hear all kinds of creaking noises. Intrigued, they turn around this group, whispering incomprehensible words. The reverend fills in his notebooks. He describes the electric murmur of the wind, the sounds of the water carving the walls, the countless plays of light. The show reinforces his conviction that nature is not monochrome but 'polychrome'. Church, for his part, paints one gouache after another with a precision that belies the low swell.

Icebergs are multifaceted. They are always changing their appearance. So much so that Noble has the feeling that he's seeing more than one iceberg when he walks along one of them. The first two bergs of a few days earlier had already captivated him. His imagination had been fired: he had seen the tent of a nomadic people in the thinnest iceberg and the vault of a greenish marble mosque in the thickest. It was as if there were secret correspondences between deserts of ice and deserts of sand. Then the masses disappeared in silence. The narrator had not even heard a sound as they fled.[4]

Among the icebergs, Noble experiences a kind of joyful stupor, like a deep empathy with another being. It is the joy of the 'Indian' faced with a deer, the unprecedented happiness of finally finding a 'wild' world. He no longer knows which metaphor to choose. One after another, he makes out Chinese buildings, a Colosseum, the silhouette of a Greek Parthenon, a cathedral in the early Gothic style, and the ruins of an alabaster city. Icebergs are great imitators. They recapitulate the history of world architecture with disconcerting ease. The Arctic Ocean becomes an open-air art gallery, a sanctuary of human creativity. Icebergs also summarise geological history. They evoke natural landforms located in the four corners of the globe. Sometimes they resemble 'miniature alpine mountains', sometimes the eternal snows of an Andean massif that the ocean has submerged. At this point in the story, Noble assures his readers that he and his painter friend share the views of the famous geographer and naturalist Alexander von Humboldt. Humboldt had just died in Berlin. He had spent his life establishing that the 'cosmos' is unified in all its parts.

This episode with the group of icebergs changes the fate of our narrator. Nothing is really the same any more. The rest of the journey is a chaos of images. The more he crosses paths with other behemoths, the more Noble forges new ones to illustrate the encounters: a warship with pointed cannons and a sharp bow, ivory carvings, clouds depicting the faces of poets, philosophers or polar bears. He describes caves, niches, balconies and escarpments. He guesses that the icebergs cast a melancholy gaze on the ship's passengers. He is saddened by the way some are obviously fragile. Meanwhile, on deck, Church finishes his preparatory oil studies. Then, in his cabin, he pencils a few sketches on the pages of a small notebook and carefully arranges his boxes.

Framing icebergs

Two years after their return, the painter unveils an impressive work to the New York public: *The North*. The painting is 1.64 m by 2.85 m. It is April 1861, opinion was positive, but not unanimous: too much emptiness, no signs of humans. Church reworked his large canvas. He decided, on the spur of the moment, to show it in Europe. In June 1863, an evening for the launching was organised in London. A number of prominent people attended, including Lady Franklin and Sir Francis Leopold McClintock. Observers in the British capital could see a broken mast, still with its masthead, pointing to a boulder on the right. Church added the detail in the final version. No doubt to evoke the tragic sinking of the Franklin and as a reply to his critics. All around the icebergs there is the same veiled Arctic glow. The painter renamed his work with the title it still bears: *The Icebergs*.

What reading can we give this painting?

A text printed on a sheet of paper was distributed when it was presented, in 1862, at the Athenaeum in Boston. In it, the artist explains his choice of perspective. He addresses the audience:

> The spectator is supposed to be standing on the ice, in a bay of the berg. The several masses are parts of one immense iceberg. Imagine an amphitheatre, upon the lower steps of which you stand, and see the icy foreground at your feet, and gaze upon the surrounding masses, all uniting in one beneath the surface of the sea. To the left is steep, overhanging, precipitous ice; to the right is a part of the upper surface of the berg. To that succeeds a inner gorge, running up between alpine peaks. In front is the main portion of the berg, exhibiting ice architecture in its vaster proportions. Thus the beholder has around him the manifold forms of the huge

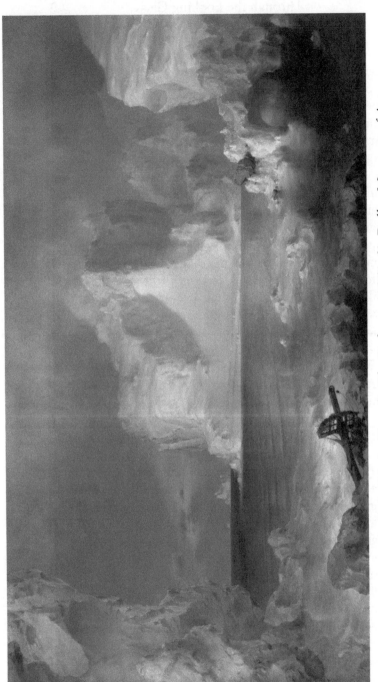

Figure 1 Frederic Edwin Church, *The Icebergs* (1861–3), Dallas Museum of Art

Greenland glacier after it has been launched upon the deep, and subjected, for a time, to the action of the elements – waves and currents, sunshine and storm.

Church trains the viewer's eye by detailing aspects of the scene. He believes that the audience needs this. For at least two reasons. On the one hand, the iceberg is a spontaneously pictorial object. But the variety of its lines must be shown. Otherwise, the viewer risks becoming bored with so much uniformity. On the other hand, the beauty of the iceberg is intriguing. The proportions of the iceberg throw Archimedes' principle into doubt. The mass seems very heavy. And yet it floats! It is so light, almost airy. How can the combination of weight and weightlessness be represented?

The painter has observed the bergs closely. He knows that their plasticity is a challenge. Their straight lines intertwine and their curves overlap. The icebergs constantly alternate foregrounds and backgrounds. They compose volumes that seem eternal. Then they dissolve into the air and the ocean. The massive ice cubes metamorphose into small balls of volatile flakes.

Church wants to control these ambivalences. He directs the gaze into a well-defined space. Better still, he plays with the frame, making the ice occupy three sides of the painting. He freezes the icebergs in their materiality and makes a stationary image from an inanimate, hieratic world that is ice in every direction, except for on high, where it opens onto a horizon tinted with the sun of a peaceful late afternoon. This framing of ice by ice, saving one side for the source of the light, has only one purpose: to make the spectators understand that the real texture of icebergs is that of light. In the eyes of the painter, it is light that reshapes the forms.

A boulder can be seen on the right-hand side of the painting. This is not an aesthetic whim. The art historian Timothy Mitchell has shown that Church was

taking a stand in a scientific controversy between Louis
Agassiz and Charles Lyell from 1845 to 1860. The
debate between the two scientists centred, among other
things, on the exact nature of 'erratic' rocks and the role
of icebergs.

Agassiz defended the thesis of an ancient global glaciation in his famous *Études sur les glaciers* and several
other lectures. During a primordial 'ice age', the Earth
was covered, and the so-called erratic boulders, which
often adorn the sides of glaciers, were signs of this.
Lyell proposed another theory in his no less famous
Manual of Elementary Geology. On several field trips
he had examined many deposits on shore that came
from beached icebergs. He deduced that these alluvia
corresponded to the rocky conditions of the continents.
His opinion was that part of the Earth, including the
North American plate, had not been covered by ice
but, rather, submerged. Then as the water had receded,
the continents reappeared and the icebergs had carried
boulders from the land. The 'floating mountains' solved
the riddle of rocks scattered far from any ice mass.

Lyell's hypotheses influenced many explorers. They
looked for evidence that icebergs carried pieces of rock.
By the 1860s, however, empirical evidence confirmed
Agassiz's arguments. His rival eventually abandoned his
theories of iceberg 'rafts'.[5]

Church's painting has gone through several versions
and many variations. Long forgotten, it is now a much
prized work. It is not only a tribute to Franklin and the
strange beauty of ice, it is a nod to a geological argument about the Ice Age. Off the coast of the island of
Newfoundland in the summer of 1859, Noble saw icebergs as the monuments representing the whole world.
The artist painted his canvases imagining that these sublime masses carried the debris of a once sunken planet.

The reign of the sublime

Church went north with a mind full of books. Like most of his contemporaries, he was aware of travelogues and scientific writings. But at the forefront of his thinking were the now popular reflections of Edmund Burke and Immanuel Kant on the sublime. In these theories of the previous century, the canonical examples are those of a mountain whose snowy peak pierces the clouds, of a storm breaking, or of a storm seen from the shore. The fictional viewer experiences a paradoxical feeling of a fear of dying while remaining in safety. All five senses warn one of the risks. At the same time, one feels infinitely free. One's reason finds strength in confronting an idea of the absolutely immense, even of the unlimited. At a safe distance, one might assume that one's life is not really in danger.

These ideas were widely disseminated among the international community of polar explorers. Confronted with icebergs, everyone feels the same contradictory emotions as those described by philosophers when faced with raging waves, lightning in the sky or alpine snow. Adventurers are both terrified and amazed, overwhelmed and exhilarated. They rediscover their dual nature as sentient and spiritual beings. They feel both fragile and powerful, both mere mortals and true demiurges. In Church and Noble's time, the spectacle of the iceberg was already over-coded by theories of the sublime.

Another argument recurs in this interpretive framework: titanic forms bear the imprint of a higher principle. Kane described the Arctic as 'a landscape such as Milton or Dante might imagine – inorganic, desolate, mysterious'. He was careful to add: 'I have come down from deck with the feelings of a man who has looked upon a world unfinished by the hand of its Creator.' The spectacular appearance of the icebergs is a reminder of the humble condition of humans. The drifting boulder is 'one of God's own buildings, preaching its lessons of humility to the miniature structures of man.'[6] The theme is an old one. Long before the beginnings of polar exploration, the iceberg was seen as a work of providence.

We are in Ireland, in the sixth century.

A monk tells another monk about his voyage in a stone trough, as in the legends of the Breton saints. He evokes a remote, isolated and magical island. All shrouded in fog, it is hidden from inquisitive eyes and unvisited by storms. On this land, nothing happens as elsewhere: nature is luxuriant, time slows down, no one feels any material need. Faced with so many marvels recounted by his own godson, Mernoc the monk, Brendan de Clonfert decides to make the same journey. After months of meticulous preparations, he and his fourteen companions embark on a small boat made of wood and leather. They set off towards the Northwest in search of paradise.

In their frail *curach*, the pilgrims huddle around the single mast. The voyage is full of hazards and miracles. They come across birds singing divine hymns and drink sleeping potions. They inadvertently cook meat on the backs of gigantic dormant fish and are attacked by sea monsters.

One day, a crystal pillar appears. It seems very close, yet it takes them three days to reach it. When Brendan looks up, he cannot see the top of the transparent pillar, which is lost in the sky. Gradually, other pillars appear. At the very top, a huge platter sits on four square legs. A sheet with thick, undulating mesh as far as the eye can see envelops it. The monks think they are looking at an altar and a tapestry. They tell themselves that this is the Lord's work.

Brendan notices a gap where their boat could slip through. He orders his companions to lower the sail and mast. They gently row into the crevice and find themselves inside a huge reticular mass. Corridors stretch out endlessly. The colours in the walls are shimmering, changing from green to blue. Shades of silver sparkle. Their fingertips graze a material that feels like marble. At the bottom of the water, they see the ground on which the diaphanous block rests. The sun is reflected in it. There is bright light, inside and out. Brendan takes measurements. For four days they calculate the dimensions of the sides. The whole thing is several kilometres and as long as it is wide. The pilgrims can't believe it.

The next day, they discover a flared bowl and a golden plate adorning the edge of one of the pillars. Brendan is not surprised. He places the chalice and paten before him and begins to celebrate the Eucharist. When the ceremony is over, he and his companions hoist the mast and sail. They take hold of the oars and set off. On their return journey, they are carried by favourable winds that bring them home without incident.[7]

Brendan's epic was copied hundreds of times between the ninth and the thirteenth century, a real bestseller. Today's commentators believe that the travelling monks saw an iceberg floating off the coast of Iceland. They would have entered the straits where icebergs calve off the coastal glaciers of Greenland.

At the time Church and Noble were travelling, icebergs were, in the European imagination, sometimes the ancestors of a geological age, sometimes creatures in the service of a sacred history. In both cases, they are icons of the sublime. Our narrator and the painter are not the only ones to see 'floating mountains' in the ocean, or cathedrals, ruins of lost cities, winding avenues, and sometimes even the face of the Creator. When ships are icebound, there is plenty of time to observe the landscape, and at such moments the romantic mind opens its toolbox and chooses the most expressive aides.

Thomas M'Keevor served in 1812 as the physician for the Selkirk settlers in the Red River Colony in Canada. In a short travelogue, he expresses his fascination for the icebergs adorning Hudson Bay. Some of them, he wrote,

> bear a very close resemblance to an ancient abbey with arched doors and windows, and all the rich embroidery of the Gothic style of architecture; while others assume the appearance of a Grecian temple supported by round massive columns of an azure hue, which at a distance looked like the purest mountain granite . . . The spray of the ocean, which dashes against these mountains, freezes into an infinite variety of forms and gives to the spectator ideal towers, streets, churches, steeples, and in fact every shape which the most romantic imagination could picture to itself.[8]

This description is already in the style of Louis Legrand Noble! It shows that icebergs have been perceived, in the Western world, as a real production in the amphitheatre of the most unbridled reveries. We all have the faculty of imagination in common. The five Labrador Inuit shown around London by Captain George Cartwright in 1772 thought St Paul's Cathedral was a mountain. They mistook the bridge over the Thames for some kind of stone structure. Those from Avannaa who landed with the explorer Robert Peary in New York in 1887 were struck by the resemblance of the first skyscrapers in Manhattan to icebergs. Each, in its own way, mixes 'the natural and the architectural'.[9] The metaphorical gaze, transposing one environment into another, is no less a characteristic of the Romantic spirit. Western travellers are wordsmiths. They make sentences. They chase the wildest of associations and spend their time imagining something other than what they see. They enter imaginary palaces vicariously and experience feelings of grandeur. It is as if they have promised to bring back postcards from these strange worlds.

Lonely spectres

In the wake of Franklin's tragedy, the far North and then the far South became obsessions that ran throughout the second half of the nineteenth century until the 1920s. Everyone had unique experiences in these icy latitudes. They are sometimes exhilarating, but always exhausting. And the closer one gets to the poles, the more dramatic they become.

Before embarkation, the explorers are full of enthusiasm. They dream of sea ice and icebergs. Roald Amundsen decided to follow in Sir John Franklin's footsteps because he had spent his nights, as a child, trying to find the Northwest Passage. Every morning, on awakening, he steeled himself to endure any hardship. His colleague Ernest Henry Shackleton tells of his dream in which he sees himself, at the age of twenty-two, on the deck of a ship in the Atlantic, his eyes in thrall to the snow and ice. His sole aim is to reach one of the planetary poles.[10] Frank Worsley imagines that he is sailing among drifting icebergs on Burlington Street, London, where Shackleton had set up an office to audition candidates for the position of captain of the *Endurance*. Next thing, he was recruited.

These desires to conquer knew no bounds. But the enthusiasm of initial dreams does not last. Reality is a quite different thing. In his 2012 *Atlas of an Anxious Man*, the Austrian writer Christoph Ransmayr recounts how he discovered the Arctic territories twenty years after writing the novel *The Terrors of Ice and Darkness*. Yet it is difficult to find a better description of the torments endured by polar explorers.

The text combines the records of the *Austro-Hungarian Imperial and Royal Arctic Expedition of 1872–1874* with the fictional tribulations of Josef Mazzini, a young man who retraces the journey many years later. Ransmayr relies heavily on the diary of

Julius von Payer, a cartographer and a first-class ensign who was also the land-based commander of the expedition that set out to find the Northeast Passage.

The crew left the German locks at Geestemünde on 13 June 1872 on the *Admiral Tegetthoff*, a three-masted schooner, heading for the North Cape and then the Pole. While on the train heading to the port of embarkation, they thought they were bound to discover an island beyond the frozen desert. Sitting in comfort, they imagined green valleys, wild reindeer, a world of freedom and life without cares. So, when they finally get there, their disillusionment will be even greater. Payer notes at the outset that

> an indescribable solitude lies over these snow-clad mountains . . . When ebb and flood do not lift the groaning and straining drift ice, when the sighing wind is not brushing across the stony chinks, the stillness of death lies upon the ghostly pale landscape. People speak of the solemn silence of the forest, of the desert, even of a city wrapped in night. But what a silence lies over such a land and its cold glaciered mountains lost in impenetrable, vaporous distances – its very existence must remain, so it seems, a mystery for all time . . . A man dies at the North Pole, alone, fades like a will-o'-the-wisp, while a simple sailor lifts the keen and a grace of ice and stones waits for him outside.[11]

The ship is soon frozen in the ice. It is transformed into a ghastly cabin. Their ordeal begins. For months they live in a 'a world totally alien', threatened by erosive activity that constantly refashions the snow. Silence is short-lived. The noise of the ice floe becomes permanent. Even its more delicate variations obsess them. Sometimes icebergs 'break with a burst of self-destructive thunder under the glow of the sun's rays . . .'; sometimes 'ice dies with a hiss like a flame.'[12]

The sailors struggle against the pressures that open up dangerous cracks around their ship. They battle against these wide gaps with trunks of oak from the ship, which 'turns, sinks and rises' in the chaos of the ice particles. They experience periods of whiteness during which there are no forms, no contrasts, no objects and no colours. The ice absorbs everything. Their exchanges are reduced to a few inaudible monosyllables and mumblings.

They keep themselves fit as much as they can, with a strict discipline. They have a clear schedule which prescribes regular physical exercise. They read a lot. But the ice does not loosen its grip. They decide to join a floe that is drifting. During their journey they come across a 'massive, rubble-covered iceberg'. The moment is sublime because 'these were the first stones and boulders we had seen in a long time, limestone and argillaceous schist.' For the sailors, these are 'emissaries' from a nearby land. They pick up debris and feel nostalgic for paradises lost as the iceberg disappears into the mist. A few days later, on 30 August 1873, a shoreline appears, and they name it after their king, Franz Josef.[13]

One passage in Payer's diary is very moving. It is the one where he lays bare their collective disappointment. One day, in a brief moment of insight, the crew members realised that 'the "North Pole" [is] not a country, not an empire to be conquered, nothing but lines intersecting at a point, nothing of which can be seen in reality!'[14] Many felt this way in the second half of the nineteenth century. North Pole explorers were coveting an invisible place. They tried to locate it with the help of cartography and deductive logic. But the absurd competition between nations turned their desire for adventure into a chimera.

What value can a goal have when one can't see it? Why reach for a geometric, purely abstract goal when snowy winds sweep across the ice, temperatures turn the body's extremities blue, one's own breath freezes, and companions give up the ghost one after the other? No one is fooled.

There is great power in these utopias of the far North and the far South. The years pass, the expeditions multiply, and so do the disasters. Despite some success, with each surge of national pride, catastrophes follow in their wake. Journeys to both ends of the world are transformed into funeral processions. The hulls of sailing ships may be lined and reinforced, but they are sheared with the crushing force of pieces of pack ice; terrible shipwrecks are the result. When they do escape, the ships are immobilised in the pack. Ice saws and gunpowder are useless. The vice is still closing. The officers find that their uniforms are unsuitable for the cold, that their bodies are rapidly weakening, and that they have not brought enough food with them. How can one resist such a hostile environment? With a few exceptions, the indigenous arts of survival are completely unknown to them. Entire crews leave and never come back. In European capitals docks are crowded with people awaiting their return in vain. Newspapers make headlines out of every drama the adventurers have. They increase their print runs. Sales skyrocket. Readers are hooked on sensation. Meanwhile, the pack becomes a land of ghosts.

The higher the latitude of the ships, the more the ice changes its form. Metaphors are of no use in estimating their volume or for describing their admirable variety. Icebergs can take on the features of tragic characters. They transcribe spectral atmospheres, not in glorious solitude but with the anguish of isolation. The ice tests the explorers' nervous systems. In their notebooks, they write down their inner states. They repeat the same phrases, use the same words, experience the same feelings. When the imagination luxuriates, it is a surface effect.

They all tell their life stories. They explain the reasons that drove them to these fractured lands. They try to justify their mission's failure and the collapse of their dreams. Silent remembrances rise from the depths of memory. The sailors note down their bizarre dreams, the hallucinations that have afflicted them, and other experiences that they have had of a supernatural character, such as yeti-like figures coming out of the fog. They are spectres, not 'rational actors in a wild region'.[15] Those who return to safety rarely return to a normal life. They become famous. But most of them are broken. The weaknesses of heroism come to light. The ice is the mausoleum of conquering nations.

A story about skulls

The young Arthur Conan Doyle is in his third year of medical school in Edinburgh. For a taste of adventure, he breaks off his studies and embarks for a few months on the whaling ship *Hope*. From February to August 1880, he sails between Greenland, the Faroe Islands and Spitsbergen. He is horrified to discover what the whale hunters did. He describes the thick fogs, and several times he falls into the cold waters. Luckily, he escapes. The experience leaves its mark. In his diary, he writes about the 'other-world feeling' that he experienced from the very first moments, and which never left him. This other-worldly feeling haunted him with the power of an obsession.[16] The budding writer did not return unscathed from his journey beyond the Arctic Circle.

In polar environments, perhaps more than anywhere else, metaphors are used as a means of orientation in an uncertain context. The captains use them, of course, to decorate the notes in their logbooks. But deep down they know that they are lifelines that help sailors to keep their minds on the job. The eye does not capture everything in front of icebergs drifting in the ocean. Such volumes seem immovable. Yet they are versatile and fleeting. Icebergs weaken the consciousness as soon as they appear. They thwart one's intentions because they do not correspond to objects that are easy to identify. The philosopher Hans Blumenberg writes that, in everyday life, metaphor generally expresses 'the need to incorporate, as part of the total causal system, even the most surprising event, bordering on a supposed "miracle".'[17]

An iceberg emerging from the mist has all the features of a miracle.

It is inexplicable: how can one describe it? The appearance of a mass on the water, the memory of the moment when it broke away from a glacier, its immense dimensions, the fear of a collision, all these elements 'disturb' one's habits. The mind feels the need to find clearer and more stable meanings. It must be able to fill its confusion. Metaphors help it to rationalise its 'situation of need'.[18] They domesticate the feeling of 'the uncanny', as Sigmund Freud put it, that icebergs inspire. They tame the irrational part of the sublime feeling, when nature is frightening. Their role is to re-establish, as far as possible, an unequivocal, or at least a more reassuring, order of understanding. For this reason, the iceberg gives rise to many metaphors. It also becomes the ideal partner for inner conversations. Everyone recognises their own image in it. On the condition that we admit its finitude. But in what way exactly?

Walter Benjamin closely studied the use of allegory in Baroque drama. He interprets this figure as the language of the creature separated from its origin by the original sin. No transcendence governs its existence any more. No salvation is possible. Everything is frozen. With the allegory, history itself becomes crystallised. The 'observer is confronted with ... a petrified primordial landscape.'[19] In the history of art, baroque allegories do not present themselves with harmonious lines, as symbols of eternity. They show skulls. The skull is the face of time.

The polar romantic spirit extends and interprets the content of baroque allegory in its own way. An iceberg floating alone carries the spleen of the fragile human condition. In the nineteenth century, many explorers and less adventurous travellers internalised this feeling. Their metaphors reflect this. Icebergs become gigantic skulls. For the fallen conquerors, these blocks are princely figures in an archaic setting. They illustrate the vanity of mankind, like personal tombs that anticipate a collective disaster. They themselves are destined to be destroyed and are glaciers in mourning. They bear witness to the glacier's slow death. As they calve, the fragments reveal the meaning of the whole. They express the evolution of a world on the threshold of catastrophe.

In the end, icebergs are fascinating because they embody a series of insoluble paradoxes. They give rise to hope as much as to despair. Their solid forms say, first of all, that they will last. They arouse a desire for permanence. Then they erode, showing that they are destined to fall apart. The block is ephemeral. On the one hand, it sculpts itself, changing its appearance in a short time. Its ability to reinvent itself is captivating. On the other hand, it reminds us of the fleeting nature of existence, impeding any real consolation.

The romantic spirit is sometimes euphoric, sometimes inconsolable. It combines exaltation and sadness, the passion for the grandiose and the allegorically natural. It admires monumental nature and contemplates death. It always sees icebergs as something other than what they are: contours and proportions that evoke familiar faces, animal profiles, the silhouettes of buildings, or even skulls. In the middle of the polar oceans, everyone interprets their morphological diversity through the prism of their desires or concerns. Sometimes they know that the causes of their changing aesthetics lie elsewhere: winds, currents, collisions, warm waters. Perhaps they even suspect that without ice there would be no humanity, or any other life.

But icebergs capture all the metaphors and reflect them back without absorbing them. They are the mirrors of personal histories as much as of myths of conquest. They exacerbate the constantly self-reflective sensibility of the Moderns.[20] The aesthetics of the sublime is an aesthetics of humans who speak to other humans.

Mirror, my beautiful mirror

Church and Noble's journey is over.

More than a century later, a professional sailor is sailing in the Southern Ocean. His mind is made up: he leaves the first run of the solo round-the-world race in March 1968. It didn't matter that he had a good chance of winning. He left the competition because he wanted to follow his own path, away from 'civilisation' and all its fakery. Using a home-made slingshot, he fires a brief message onto the deck of an anchored oil tanker: 'I am continuing non-stop to the Pacific Islands, because I am happy at sea, and perhaps also to save my soul.'[21] Despite his melancholy, he moves even further away from his family and friends and ventures towards Cape Horn and the Pacific, towards the Galápagos Islands.

Bernard Moitessier listens to the wind during his flight. He talks to the clouds and counts the raindrops on the sails. Dolphins accompany him for a few days and guide him off the reefs. Childhood memories come back to him. He cuts his beard every week to enjoy his own morning porridge clean-shaven and does yoga exercises in the cockpit of his small sailing boat. He travels towards peace and freedom. He lives in the present in order to 'forget the world, its merciless rhythm of life'.[22]

The sailor fears that icebergs will cut his path. His nights become more chaotic than usual. He hardly sleeps at all. When he sees flocks of Cape larks flying around in the distance, he suspects that he is about to enter an area of coastal ice. He is wrong and manages to get around the drifting blocks. He writes: 'Seeing an iceberg in fine sunny weather. It must be the most beautiful sight a sailor could lay eyes on, a thousand ton diamond set on the sea, glittering beneath the southern sun. It might be enough to last me the rest of my life.'[23]

Unlike James Cook, Moitessier did not cross the Antarctic Circle several times in an attempt to discover the continent that scholars and ministers talked about in the gilded halls of the British capital. He is not brought to a stop before an impassable wall, peopled by giants. When he returns and publishes his story, the collective imagination does not take off like a grass fire. The world is not enthralled by the 'islands of ice' floating in the Southern Ocean. His 'diamond' sparkles are not as mystical as those in Brendan de Clonfert. The allegorical tendency of polar romanticism is foreign to him. Unlike Noble, he is not one for rhetorical emphasis. Instead, he relishes simple words that speak of the possibility of a quite profane ecstasy.

He is afraid of icebergs. But he assumes that seeing just one of them will satisfy his wanderlust. He knows that many have wanted to pass through an imaginary mirror and have bumped into their own faces, like Narcissus in the clear water of a spring. Some have perished. He just runs away.

The iceberg is the mirror that the Moderns set up between them and nature to contemplate themselves. They imagine another side in order to get closer to their reflection. Self-esteem needs a magnetic North or South, a specular material.

But, over there, the cold does not become hot, the maps of the world do not invert, the compasses do not come to life like the chess pieces in Lewis Carroll's tale of Alice's adventures in Wonderland.

There is no mirror there.

2

The Eye of the Glacier

The night was becoming increasingly opaque. With headlamps adjusted to foreheads, we progressed in single file along the steep mountainside. The path was still well marked, the lights reassuring. The valley had disappeared below. In front of us, immense proportions. The higher we climbed, the smaller we became.

We had left just before dusk. We had been walking for a few hours and our field of vision was narrowing. I was dependent on the shaking cone of light in front of me. All around our procession, creatures that hide during the day were awakening. The owls began to hoot. Perhaps they were looking for dead wood to build their nests in a trunk that a woodpecker had abandoned. Perhaps they were males singing nuptial songs. They brushed past us in silence. Without doubt they were watching us with their big hypnotic eyes, like revolving headlights.

Beyond the tree line, the path became lined with small piles of pebbles and rocks. We were entering the moraines. We had to attach our ropes to avoid any unnecessary risk.

So other, so close

As far as I can remember, the hut appeared a little before 2 o'clock in the morning. It was discreetly perched on a rocky promontory with no more available space. We checked out the terrace. It was quite welcoming, despite the bitter cold. As the sky was starry, we decided to have a rest outside. An hour of sleep, no more. We had to get back on the trail before the snow melted further up. All around the hut, ice had replaced the moraine. Once the crampons had been attached and the harnesses put on, our company set off again, rope taut and with a heavy step. A long silence settled in. All we could hear was breathing. We had the summit of Mont Blanc du Tacul in our sights.

Suddenly, the view closed up. It was very dark around us. My headlamp flickered, as if the ice floor had started to shake, and the gradient sloped even more. One of the leather straps on my crampons was probably loose. My shoe slipped, my ankle twisted. I stumbled and lost my balance. Below me, lights flashed: my ice axe disappeared into the darkness of the valley. My right leg had hit the rock. My knee was bruised. I, for one, would miss out on the summit.

Many years later, we are driving along the bed of an ancient glacier, now a stony dirt road. Our friend Þorvarður Árnason is concentrating. The mud fills the wheels of his jeep, and the uneven terrain throws us in all directions. A fine drizzle has been falling since the morning. High walls surround us, as if in a valley floor. Then we abandon the vehicle and start the walk that should lead us to the foot of the glacier. Not far away, we can hear the rustle of a waterfall. We cross a field of boulders piled up on top of each other. Not a tree in sight. There are only rocks of varying sizes.

We see it at the foot of a hill. The ice is surrounded

by a belt of clayey silt. We descend to a sort of stony clearing and enter a cave at the bottom. The sun is weak but its rays pierce the damp curtain. They illuminate the transparent ceiling. Frozen shapes appear: air bubbles, filaments of dust, sprays of grass and leaves, fragments of flint. Everything is crystallised. The smooth walls of the cave are in shades of blue. We caress the silhouettes. The hand does not stay long. At the bottom, a current sprays onto the rock. The torrent roars. Above, the glacier creaks. It perspires as well. Drops fall to the ground.

We crawl on all fours so as not to touch the ice stalactites. I learn that they are yesterday's and that they will not last the night. Dusk arrives. We exit again. The place looks like a lunar outpost. It is completely different and asks nothing of us. We say goodbye to Breiðamerkurjökull, as if to an old acquaintance.

From the tangle of seracs to the cave of a formerly coastal glacier, the memory of my aborted mountain trek reflected back on our hike that day. In heights of the Alps I had already had the feeling that the ice had moved under my feet and that I had awakened a being that was languishing in its sleep. As if we had entered a place where we were making too much noise and its owner had watched us with a frown. In the southeast of Iceland, I again had the impression that an eye was watching us, that a spirit was spying on our movements, listening to our conversations and enjoying our silences. I was as fascinated as young Unn in Tarjei Vesaas's *Ice Palace* when she guesses that she is not alone in the crevices of a great waterfall that has been transformed by the ice into a marvellous mansion.

Several memories came back to me. They were white impressions: the fine, crunchy snow in the woods of my childhood in Touraine, a landscape of well-powdered fir trees in the Black Forest, the sound of dripping under a bridge of frost at the bottom of a valley in the Queyras, a blinding light during the brief stop of a train in a high-altitude station between Oslo and Bergen.

These winter memories had already resurfaced. This time they made me even more perplexed by the modern need to place a mirror between oneself and nature. I had been intrigued for some time by the endurance of so many travellers who stumbled, like innocent insects, on their reflections, believing themselves protected by their utopias as much as by their fears. In Iceland, I understood that some of them had felt scrutinised.

They lacked the words to translate this experience. They were trying to solve an enigma that was not just one of a sublime spectacle. It was that of a different life that had to be appreciated for itself. They wanted to see those who were watching them and who were not always pleased. Since they lacked the appropriate language, their best guess was that the mirror was one-way.

Way up above

A drop of liquid falls from the sky in the shape of a snowflake.

The writer Eugène Rambert reminds us that snowflakes can travel great distances, from the Atlantic Ocean to the peaks of the Alps and back to the sea via the Danube, the Rhone, the Po or even the Rhine. Its wandering lasts from a few hours to half a century. It all depends on the winds. Then it reaches the ground. Its needles become rounder and eventually disappear. The flake becomes an ice crystal. It merges with other flakes that do the same. A *névé* is formed. The crystals agglomerate again and become a glacier. In this way the snow piles up into compact ice on the slopes of the Bossons glacier in the Chamonix Valley. Then the crystals melt and the flake returns to its original condition as a drop of water: 'Describing a glacier means telling the story of this journey.'[1]

The snowflake's adventure is experienced differently in the mountains. The winter months were for a long time a period of anxiety and withdrawal. They were dreaded because it was a time of blizzards, snowdrifts and travel that was frustrated, sometimes made impossible. Bells warning of storms helped those who had lost the trail to reach the village.[2]

Snow was associated with the disasters caused by glaciers breaking out of their beds. No prayer or procession of monks could stop them wreaking havoc – 'surging', as the Anglo-Saxon glaciologists say. The internal water pockets, which form during the summer melt, gave way and flooded first the pastures and then the habitations below. Blocks of ice shot up into the air and fell in pieces on the houses. Lakes with small icebergs floating in them appeared. Entire valleys were blocked by serac avalanches. Ice avalanches are part of the history of the Alps. Memories of monumental ice break-ups were a feature of fireside chats many years later. The fear of such events made mountain people stick together. The villagers protected themselves together from the wrath of the hanging behemoths. Together they rebuilt their walls and streets.

The glaciers 'up there' evoke a distant world. Sometimes we avoid it. Sometimes it meets us. Crossing snow-covered mountain pastures is never without risk. When Élisée Reclus became interested in this natural environment, winter travel still required residents to take great precautions. Most vehicles were not able to withstand winter's fury. Sledges replaced carts and mules to cross the passes. They offered privileged views of the relief and veins of the landscape:

> It is by travelling in this way by sledge over the mountain passes that one can learn to get to know the great

snows well. The light frame glides noiselessly; you no longer feel the impact of the iron on the solid ground, and you feel as if you were travelling in space, carried away like a spirit. Sometimes one skirts the curve of a ravine, sometimes the projection of a promontory; one passes from the bottom of chasms to the edge of precipices and, in all these varied forms which one views one after the other, the mountain maintains its uniform whiteness. If the sun shines on the surface of the snow, innumerable diamonds shine; if the sky is grey and low, the elements seem to merge. Clouds, snowy mounds – all look the same. It is like floating in infinite space, no longer belonging to the earth.[3]

Travelling by sled was a way of incorporating the countryside, of living at its pace by following its contours. Close to the snow, one understood its varieties, circumvented its traps, and moved with the ease of a bird in flight or a high-country sprite. It was on the way back down that the mountain would test you most rudely, especially during the touring season. The sled often got out of control and plunged down the slopes. It gave the impression to those driving it that it was causing an avalanche because it carried so much snow in its wake. In this 'infernal gallop', this 'frantic race', the promontory or the ravine, hidden under the snow, were the worst enemies. The arrival at the foot of the mountain was experienced as a relief, sometimes even as the end of a long 'hallucination'.[4]

Reclus hesitates when describing the glacier at the very top of the mountains. In contrast to the 'joyful world' of nature below, he describes it as a vast environment, 'dull, with its gaping crevasses, its piles of stones, its terrible silence, its apparent immobility. It is death abutting life.' He uses the image of a 'shroud' to evoke the snow that covers the contours of the ground, masking its difficulties and chasms.[5] But he knows that the movements of the glacier shape the planet's surface 'with invincible force'. Its snows return to the clouds as they evaporate. Tiny molecules fall and make ice again. The glacier unites earth and sky. It is part of the cycle of life.

The geographer classifies it in the register of 'inanimate nature'. However, he uses terms that give it a soul, especially as spring approaches:

> When one has become intimate with the glacier through long explorations, and when one knows how to become aware of all the little changes that take place on its surface, it is a joy, a delight, to walk along it on a beautiful summer day. The warmth of the sun has given it back its movement and its voice. Veins of water, almost imperceptible at first, form here and there, then unite in glittering rivulets that meander along the bottom of miniature river beds they have just dug for themselves and suddenly disappear in a crack in the ice with a small, silvery-voiced wail.[6]

As if by magic, the sun heats the air. It transforms the streams into torrents that drain the alluvium onto the banks. With it, the glacier takes on a new voice. It is literally resurrected. The essential point is this: the glacier moves and expresses itself. Better still: it 'continually walks', its 'windowed mass vibrates with a continual shudder [and] shakes the snowy mantle that covers it.'[7] It is as if it were the den of an animal stretching out its limbs after a long period of hibernation and preparing to leave in search of food.

In the end, Reclus acknowledges the results of the research carried out by nineteenth-century glaciologists. Hypotheses on the mobility of glaciers could vary (the dilation of the crystals, their refreezing, water at the base which makes them slide); Louis Agassiz, John Tyndall or James Forbes reached some simple conclusions. Firstly, the centre of a glacier moves faster than its sides. Secondly, its lower part and lower reaches move more slowly than its upper part and upper reaches. In a meandering glacier, however, the side towards the outer curve moves faster. Finally, a glacier moves more slowly in winter than in summer.[8]

The scientists were not unaware that these laws of mobility could seem paradoxical; at first glance, glaciers are immobile. They are so ancient, so permanent, so fixed. But their laws proved the opposite, and that was the important point. In 1840, Agassiz summed up his thesis about the high Swiss glaciers when he wrote that 'everything here, in a word, bears the imprint of mobility and movement under the guise of immobility.'[9] The villagers were right to fear 'glacial surges', the sudden advance of masses of water down the mountain slopes.

However, one question remained unanswered: if they move, are they alive?

A vocabulary crisis

Whatever one thinks of any of the attitudes mentioned so far, one can perhaps guess that certain assumptions must be abandoned. But one question keeps coming back: can we see the world as it is, without mirrors, and recognise the full variety of its life forms? This is not, on reflection, an unreasonable claim. Let us say that the contexts still need to be specified. Let us focus on the events that suggest to their observers that glaciers and icebergs have personalities.

Louis Legrand Noble and Frederic Church had witnessed the internal break-up of an iceberg during their journey off the coast of Labrador. They had narrowly avoided the cataract of ice that collapsed on itself not far from their boat. Once over this scare, the narrator remembered a conversation he had had with the consul and a member of his church during a dinner on land.

The latter had told him that the breaking of a glacier was the most beautiful thing he had ever seen in his life. He had felt that the iceberg that had broken off had joined an invisible avenue and that it had moved gloriously towards its end. He compared this phenomenon to an 'earthquake'.

Once on the deck of the schooner, Noble learns from their experience with the crumbling iceberg. First, he understands that a loss of mass at the surface creates an imbalance and causes the whole thing to move in on itself, as when you stumble and want to regain your balance. He distinguishes the 'iceberg of the air' from the 'iceberg of the deep sea'. He then correctly notes that the life of the whole mass depends less on its movements visible to the naked eye than on its movements below the water line 'upon its centre – its rotation and vibration'.[10]

Noble continues to reflect on the disintegration of the iceberg. The blocks give him the impression that they wander on the water like white clouds in the sky. They actually float 'just *under* the surface of the ocean'. When they break, their driving force is 'explosive power'. They detonate like gunpowder going off, with a thunderous sound. The sound of their break-up is terrible. It fills the whole atmosphere. Even a pine tree struck by lightning does not inspire such fear! Then we return to the moment when the iceberg calves from the glacier. It was already an explosion-like voice before the 'final plunge' into the frantic waves. The spectacle of a cathedral collapsing in on itself or of a mountain opening up would be of the same order. The reverend can't help but hear the organ of the Apocalypse playing, the work of the Creator who always creates and destroys. He imagines the 'obelisk' embarking on its final journey, amid a crowd of ruins in the foamy waters. The mass is condemned to turn on itself, getting thinner and thinner, until it is all gone.[11]

These visual descriptions occupy the end of a chapter entitled 'The Story of an Iceberg'. At the beginning of the book, Noble said that it was without doubt 'the most wonderful story that can be written'. It is an important chapter. In it, the narrator enumerates the achievements of his text and sets himself up as a biographer of a material being. He observes at this precise moment that 'icebergs, to the imaginative soul, have a kind of individuality and life.' He thus explains himself while remaining faithful to the rules of his rhetorical art. As usual, he provokes associations between all forms of the world: the bergs are sometimes clouds, mountains, monuments, sometimes angels, demons or animals; they combine the lines, surfaces, movements and colours of the sky, earth and water. This is why they 'startle, frighten and awe; they astonish, excite, amuse, delight and fascinate.' But Noble changes his tune and adds, more concisely this time, that icebergs have 'voices': they speak a 'living language' to those who listen. We think they look the same. Yet each one 'differs widely from all others' in its sounds, its movements and even its choices.[12]

The change of tone is worth taking seriously. It signals two orders of language at work in the story. One is emphatic and grandiloquent. The other is restrained and more sober. Four chapters earlier, the hypothesis of iceberg souls had already been mentioned. The vocabulary has been stripped away. It is as if the assumption that icebergs live and have distinct personalities has created a disturbance in the discourse.

The writer is not always comfortable. His imagination wavers. Sometimes words no longer help him describe what he hears. He hesitates and summons his last strength. Common metaphors are useless. He goes through a real crisis of verbalisation: 'I am quite tired of the words: emerald, pea-green, pearl, sea-shells, crystal, porcelain and sapphire, ivory, marble and alabaster, snowy and rosy, Alps, cathedrals, towers, pinnacles, domes and spires. I could fling them all, at this moment, upon a large descriptive fire, and the blaze would not be sufficiently brilliant to light the mere reader to the scene.'[13]

In the midst of an image euphoria, and while accompanying a renowned painter, Noble admits that substituting memories of European monuments or images of Eastern landscapes for icebergs is not always appropriate. The masses are not incidental details in a world for which we are the sole interpreters. Their sonic truth exceeds the powers of the imagination as much as those of his fellow traveller's paintings. No theory of the sublime can account for this truth. The reason is simple: the centre of icebergs lies below the waterline, in the realm of the invisible. Sublime aesthetics, on the other hand, sticks to the realm of the visible. It is only interested in the 'iceberg of the air'. When the berg rolls over, the submerged part of it noisily emerges: its soul appears. This kind of event is more intense than any language. It does not need a spectator.

The iceberg thus calls into question the ability to decipher the world. It can even render poetic language useless. The narrator confesses his dismay and defeat. He hopes that the Icelandic or Greenlandic languages have more suitable terms than his own. But he doesn't loiter; he decides to pay more attention to the material and its intonations. He refocuses his speech. As a good biographer, Noble knows the end of the story and is especially interested in climactic moments. He rediscovers his metaphorical spirit. In the chapter on the 'story of an iceberg', however, the life cycles of the bergs are described more sensitively than in the rest of the book. The sounds become significantly more important. We hear the sounds the iceberg makes when it collapses in on itself: 'crack-crack-crack'. His gestures are always monumental. But the dramatic tonalities outweigh the overall effects. The moral of the story is simple: to understand an iceberg, sounds count more than words. Acoustics are a must.

The wanderings of a happy man

Perhaps some illusions are beginning to fall away. How can we proceed now? How can we see the iceberg and the glacier as animate beings? What experiences, what categories, what attitudes are necessary? And for what purpose? One option would be simply to assert that the entities in question are sentient and go on from there. This would be going too fast. For we would have to detail the how without having explained the why. The other option is the one we choose. It consists of continuing along the same path, moving forward in small narrative sequences. Let us continue to leaf through our album of experiences for a while, telling its stories and evoking significant figures.

Between the summers of 1879 and 1899, John Muir surveyed Alaska seven times, as if to verify the vocal power of icebergs noted by Louis Legrand Noble. His passion for glaciers dated back to his student days and remained with him ever after. He was the first to believe that Yosemite Valley was carved out by glaciers. During his excursions to the far North, he was accompanied by a missionary friend from Fort Wrangell, Samuel Hall Young, along with native guides and a few other associates. At that time, the Tlingit Indians asserted that they had entered the Wrangell area through a tunnel dug under the glacier by the Stickeen River.

The naturalist boards the *Cassiar*, a river steamer. When the boat reaches the area where the glaciers are found, he talks to the passengers and answers their questions about the geological history of the area: 'Are these white masses we see in the hollows glaciers also?'

'Yes.' 'What made the hollows they are in?' 'The glaciers themselves, *just as traveling animals make their own tracks.*' 'How long have they been there?' 'Numberless centuries.'[14]

The expression rings true. Some animals manage to hide their tracks. Glaciers, on the other hand, are the easiest creatures on earth to track down. You don't have to smell them or be on the alert. You have only to look around you to see the landforms that their progress has shaped. Their furrows are deep and the age-old proportions are still visible. The snow does not erase the ancient slashes carved into valleys. The signature of the glaciers speaks for itself. When Muir saw the striations on the walls of one of them, he wrote the following thoughts in his notebook:

> Standing here, with facts so fresh and telling and held up so vividly before us, every seeing observer, not to say geologist, must readily apprehend the earth-sculpturing, landscape-making action of flowing ice. And here, too, one learns that the world, though made, is yet being made; that this is still the morning of creation; that mountains long conceived are now being born, channels traced for coming rivers, basins hollowed for lakes; that moraine soil is being ground and outspread for coming plants . . . to make the mountains and valleys and plains of other predestined landscapes, to be followed by still others in endless rhythm and beauty.[15]

Glaciers reinforce Muir's Darwinian convictions. They show that the Earth is in a state of continuous birth: where they disappear, liquid expanses, new flora, other mountains will soon appear. The process of fragmentation is neither a sign of an abandoned world nor evidence of an inevitable decay. It indicates the incessant movement of life, the play of evolution. Muir, the father of the national parks in the United States, was convinced that nature metamorphoses. It is always regenerating itself, as if it were growing old while growing young. It is somehow imperishable, even if it suffers and needs to be protected. A vital force holds all its parts together. Glaciers follow the rhythm of the centuries. They have shaped the planet for thousands of years. For this naturalist, they will not stop any time soon.

Glaciers are not landscapes. They are landscapers, sculptors who polish, grind and slice up the folds of the land over which they slide, without ever getting tired. Even when they dwindle, they continue to work away. It is obvious that they move by themselves and live in their own way. They are never still. Their advance, in any case, goes at different speeds. They can advance by thinning, retreat by thickening, or melt from within.[16]

Icebergs, on the other hand, fall into the water like avalanches. Sometimes they pivot on their submerged base and shake a great mass of water as they suddenly collapse. The physician and explorer Elisha Kent Kane, looking at the icebergs of Greenland, felt that the Creator had not completed his work and had abandoned this part of the world. When Muir observed the erosion processes that had shaped the Wrangell area, he never thought of ruins, and even less of a wasteland. His curiosity is primarily scientific: the study of an iceberg makes it possible to analyse the mass from which it comes; the fragment is as interesting as the whole; the sample has its own dynamics.

But where do icebergs come from? Many of Noble's friends thought that bergs were the result of a build-up of snow, sometimes powdery, sometimes frozen, in the Arctic Ocean regions. This snow would pile up on top of itself and be cemented together by the cold, plus new snow and sea spray, until it forms a solid island. This theory was still widely accepted in the mid-nineteenth century. When Noble objected, on the basis of scientific evidence, that icebergs were glaciers 'first formed on the land, and then launched into the sea', his listeners only shrugged, sometimes even expressed outright disdain.[17] Muir has no doubt. The iceberg is the ephemeral piece of a glacier that carries the memory of mythical journeys, from the first snowflake falling from a misty sky to an ice-covered land from which it tears itself with noise and fury.

When the clouds darken, Muir sometimes gives in to melancholy. The icebergs that line Glacier Bay are like 'a solitude of ice and snow and newborn rocks dim, dreary and mysterious'. But he is much more often amazed by the glaciers that synthesise past, present and future, symbols of a virgin nature living at its own pace, echoes of divine work itself. The naturalist surveys the Mount St Elias region at a time when others are struggling to reach the North Pole. He does not feel the need to go so far north, or even to go over to Greenland to see what so many others have seen. The spectacle of the icebergs is happening right there in front of him.

During one of his excursions, he descends into a pit. The place turns out to be a 'kettle', a cavity dug by streams resulting from the melting of a glacier. He reaches a mountain lake that resembles a 'miniature Arctic Ocean, its ice-cliffs played upon by whispering rippling wavelets and its small berg floes drifting in its currents or with the wind, or stranded here and there along its rocky moraine shore.' He puts down his bag, unfolds his tent and begins to tidy up his belongings and food. Then he gets ready to lie down to sleep. Suddenly, he hears a thunderous roar. He gets up, climbs to the top of the moraine and stops, stunned by the scene: 'the tremendous noise was only the *outcry of a newborn berg*, about fifty or sixty feet in diameter, rocking and wallowing in the waves it had raised as if enjoying its freedom after its long, grinding work as part of the glacier.'[18]

If the glacier is an animal that leaves tracks behind it, it is a living being. It is therefore natural that it should give birth and that the iceberg should be its child. The reasoning is logical. Muir describes the break-up of the icebergs in terms of giving birth. Such events do not make his talk too unruly; in Alaska, the naturalist is in better control of his metaphors than Reverend Noble off the coast of Newfoundland. He abandons his companions, without even warning them, so as not to miss one note in the symphony of birthing icebergs. Alone, he ventures onto the glaciers. He comes across 'a remnant of the glacier' that seemed to him to have been 'preserved by moraine material' and on which centuries-old trees were growing. He identifies a retreating glacier. His main objective is to explore Taku Fjord to understand how icebergs are produced by glaciers.[19]

Muir observes fault lines in these masses. He guesses that ice is a material that does not respond instantaneously. Its normal rate of transformation is slow. It retains its internal instabilities for a long time. Then a process of fracturing is triggered which alters the general balance of the glacier. The naturalist spots the zones of fragility, the old or new ruptures that run between the surface and the base. He understands that the future iceberg is detached when a crevasse at the base coincides with a crevasse at the surface and spreads to the water level. In this process, hearing assists sight.

As he slept, Muir was awakened by waves licking the borders of his tent. Produced by falling heavy icebergs, the highest of them

> oftentimes travel half a dozen miles or farther before they are much spent, producing a singularly impressive uproar in the far recesses of the mountains on *calm dark nights when all beside is still.* Far and near they tell the news *that a berg is born*, repeating their story again and again, compelling attention and reminding us of earthquake-waves that roll on for thousands of miles, taking their story from continent to continent.[20]

The lakes of mountain ranges and the coastal shores are usually calm, rocked by somewhat weak movements of water. The falling of icebergs is an event that is all the more expressive because it disrupts the regular state of the liquid sheet. So, the blocks are distinguished less by their shapes than by their sounds. They are not necessarily very massive. They have protruding edges and stratified structures that bear the memory of the snow at high altitudes. But, above all, they agitate the surfaces to the point of raising immense waves that race along chaotically: the waters carry the news that a being has just been born. In the mountains or along the coast, glaciers and waves cooperate readily. A distinct camber, deep waves, a high ridge line; the backwash created by the fracture transmits the information. The waves are the glaciers' messengers.

It is a small tsunami! The event disturbs the whole region. All living things notice it, underwater as well as on the shores and far inland. The sound is so powerful that the fauna in the forests can hear it clearly. This great disturbance is a sign of life. The icebergs speak for themselves. They do not need interpreters: they are members of nature at large.

Muir is captivated by this aspect of the life of the icebergs. He tells us icebergs are hitting each other every five to six minutes. He notes that the sounds sometimes come from the submerged part of the glacier, invisible to us. The bergs make even more noise. They bounce back to the outflow face, then plunge and plunge again until they roughly stabilise. He deduces that, if the ice on land is shrinking, the water level is generally rising.

Around the fire in the evening he asks his Tlingit guides about the respectful relationship the Indians have with the animals they hunt. He asks them what they think about the souls of wolves. But his mind is elsewhere. He spends all his days 'watching the ice-wall, berg life and behavior, etc.', of the frozen expanses.[21]

The birth of icebergs obsesses him. Figuring out how this separation works fills him with joy.

Are icebergs whale calves?

We might be tempted to avoid the language of birth. We could point out that it is evasive and confusing when applied to an inanimate being. But then we could assert two things: firstly, that we are speaking improperly – matter cannot be born; secondly, that we are widening the scope of the concepts used, including that of life. There is thus an amalgam, since the same term, birth, is used to designate living and non-living things.

To reject the language of birth would be to admit a strict distribution of roles between so-called animate and inanimate beings. Yet the boundaries between what is alive and what is not are more blurred than we think. When did we start describing the breaking off of a glacier as a birth? Perhaps unknowingly, Muir had inherited lexical practices from a specific context.

On 19 November 1857, the photographer and science writer John Thomas Towson gave a lecture to the Lancashire and Cheshire Historical Society. He explained to his learned and curious audience that icebergs have an appearance that distinguishes them from other forms of ice. Like so many of his contemporaries, he reminds the audience that the appearance of these drifting blocks can be fanciful, sometimes resembling precipices, sometimes pinnacles or chalk cliffs. The icebergs rise hundreds of metres above their waterline and glisten with eerie green colours when the sun shines onto their crevices. Sometimes you can find quite azure-blue lakes at their summit. He went on to say that

icebergs are not the produce of one season; on the contrary, there is reason to believe that these masses

commenced their formation at a period equally remote
with that of the origin of some of our tertiary rocks.
They are of the same nature as the glaciers of the warmer
regions of the earth; but instead of being melted in the
valleys, they are pressed forward into the ocean till at
length the water is sufficient to float them and immense
blocks are broken off. This process has been termed
by the Greenland whale fishermen the 'calving' of an
iceberg.[22]

The term is striking, more precise than 'birth'. Towson
is not the only one to use it. Before him, the geologist
Hinrich Johannes Rink used the same term, with a few
differences. When the fragment of the glacier falls into
the water, he wrote, the action 'is called *calving*, and
such is the concussion that it sometimes sets the sea in
motion to a distance of sixteen miles. From the above, it
is evident that icebergs cannot be considered as breaking
off from the coast; it would be more proper to say *that
they rise* out of the sea.' Rink, a pioneer in glaciology
and founder (in 1861) of the first Greenlandic language
journal, seems to favour an explanation of 'calving' by
wear and tear or breakage of the submerged parts, what
today's glaciologists call 'undercutting'. He notes that
it was the Danish settlers who 'singularly called . . . the
dropped fragments of ice . . . *calf-ice*'.[23]

As applied to glaciers and icebergs, the term 'calving' dates back to at least the first half of the nineteenth
century. The Reverend Henry Theodore Cheever recalls
that an American whaling ship, named the *Pacific*, was
lost in 1807. The captain had tried to moor the ship to
a large iceberg by holding his ship at a distance with
chains. He sent a longboat for his men to plant anchors
in the ice. S-shaped imprints were already visible on the
walls of the block. But, just as one of the sailors hit it
again, 'a crackling noise was heard and presently several
large pieces, or *calves* as they are technically called, fell

off into the sea.'[24] Then the whole mass collapsed and the ship keeled over dangerously.

It can be assumed that such a lexicon was circulating among indigenous Greenlanders at the end of the same century. The author of an article published in August 1892 in *Scribner's Magazine* said: 'the Eskimaux are so familiar with this process of separating bergs from the land-ice that, when they hear the roar which it causes, they say that the glacier "is calving" or "giving birth to its young".' Prejudiced, he adds that this description is 'savage'. It is 'common to primitive' and 'uncivilized' peoples who breathe life into rivers and ocean currents, as into all of nature. It dissipates of its own accord when culture develops and reason triumphs.[25]

Representations of the Inuit are judged here in the same way as the critique of the natural beliefs of the natives and the struggle of the civilised mind against their supposed barbarity. The philosophy of history that fuelled the politics of territorial conquest at the time is reflected in this passage of the text. Yet, at the same time, the author reveals essential aspects of the system of thought that he rejects.

One fact seems certain. All these quotations attest to a widespread use of the term 'calving' among American, Danish and Greenlandic whalers in the nineteenth century. This usage is undoubtedly linked to a general history of commercial practices. Is it even older? Did the Scandinavian settlers in Greenland inspire the Inuit when they came with their herds, or did it originate among the Inuit?[26]

Literally, 'calving' means that the glacier gives birth to an iceberg in the same way that a cow gives birth to a calf. The word was already in use for marine mammals. In the tenth edition of the taxonomy of the naturalist Carl von Linné, which dates from 1758, the Latin name for the harbour seal is *Phoca vitulina*, or 'sea calf'. It is applied to the babies. In English, the term 'whale calf' is used to designate the baby whale. In the environment of the fishermen who scoured the waters of Greenland, it is reasonable to assume that the iceberg was associated with a whale calf rather than a land calf. Sources are so scarce that it is difficult to know which language first carried this set of associations.

Scholarly etymology is important. But the key is to understand that 'calving' makes the iceberg a living being, a warm, not a cold, substance. The fishermen of the time described it in terms of activities they share with marine mammals: being born in the water and spinning around. They freed it from the romantic discourse that saw in it all the monuments of the planet. The iceberg was no longer the object of an unbridled imagination portrayed by using metaphors and analogies. 'Calving' is still a metaphor in the eyes of linguists. This figure of speech is above all an existential figuration that brings together two non-humans, the iceberg and the whale calf. At the price of a certain paradox, fishermen have brought the icebergs into the world of life. They have animated them.

Scientists quickly accepted this usage from the seafaring and fishing professions. Glaciologists use the word 'calving' to characterise the breaking processes that precipitate the terminal tongue of a glacier into the sea or into a lake, and even in the mountains (the expression 'dry calving' is used when a serac breaks off). This may be poetic licence. The term is nonetheless part of the academic lexicon. It is present in all studies of how glaciers respond to the seasonal variations that determine their fractures. Non-specialist dictionaries have recorded it. This is not really surprising; to take an interest in the sensitive phenology of these masses is to find out that the ice is 'vibrating'.[27] It reacts, like any other living being, to the cycles of the Earth.

The spirit of laws

Every organism is subject to changes of state and is regulated by multiple chains of cause and effect. Without claiming to take the place of professional glaciologists, we shall try to present clearly the phenomenon of calving and certain aspects of the life of the iceberg once it has fallen into the ocean. Then we will offer conclusions that the mind can literally draw from these physical laws.

In an ice mass, nothing is stable. There are movements compensating for each other, both at the front and at the top, and at the water level or below. The overall balance is constantly being redefined. 'Thermal melting' and 'mechanical disintegration' are the main causes of glacial break-up. The former reduces the volume of the ice, while the latter breaks it off.[28]

What exactly happens with a melt? In a nutshell: crystals evaporate due to the heat and others pile up on top of each other. Air bubbles are released. In the case of a coastal or lake glacier, winds and waves erode the walls in contact with the water. Internal cracks, ridges and fissures appear. These are points of frailty, or 'calving zones'. Some parts start to topple over as the whole thing cracks vertically. Breakage occurs if the cracks grow 'around the same perimeter'. The mechanical process is then said to 'assist' the thermal process.

It all depends on the size of the glaciers and the condition of their 'mills'. Mills are shafts that are formed at the surface. They carry meltwater to the bedrock and help the masses to slide down more or less slowly. Small glaciers melt in less time and calve smaller blocks of ice. Large 'tongues of ice' are more resilient than the small ones. Their terminal part 'lengthens' and gradually 'thins'. Land-fast ice or grounded icebergs are likely to 'delay' the calving phenomenon by limiting the action of the backwash. But the masses weaken under their own weight;

they bend more and more as the tides continue to work against them. Large chunks end up dropping into the water all at once. In general, these births occur 'between the hinge and the tip of the tongue' of the glacier.

The scientific language adapts itself to the variety of phenomena observed. It models the many stresses on the ice fronts. For the sake of simplicity, let's identify, with the help of glaciologists Douglas I. Benn and Jan A. Åström, four general 'styles' of calving (see figure 2). As may be seen, an iceberg can be formed by very different fractures. Pressure movements push the ice mass lengthwise towards the sea (A). The dynamics of deep melting cause the upper vertical surfaces to fall into the water (B). At times, a foot of ice is lifted (C). At others, an entire piece of ice tongue is released (D).

Clearly, wave vibrations play a significant role each time. They increase the 'hydrostatic pressure' on the glacier.[29] However, a glacier that breaks up does not

A: Longitudinal extension

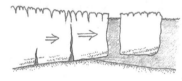

C: Buoyant calving – ice foot

B: Melt undercutting

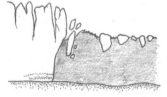

D: Buoyant calving – full thickness

Figure 2 Diagrams extracted from Douglas I. Benn and Jan A. Åström, 'Calving Glaciers and Ice Shelves', *Advances in Physics: X*, 3/1 (2018), p. 1053.
© Benn and Åström, 2018.

necessarily immediately release an iceberg. The glacier tongue acts as a 'cantilever beam'. It is sometimes supported by the 'thrust of the sea', which makes it float, and sometimes by the 'rigidity of the ice'. Sometimes the sea cannot support the volume of the ice sheet when the beam breaks. The future iceberg then 'hangs on, sometimes for decades'.

Once the iceberg is free of the glacier, it is entirely bound to its maritime environment. The swell makes it 'oscillate' if its ice is 'rigid'. It makes it 'vibrate' if its ice is 'elastic'. The movement causes its mass to emerge to a greater or lesser extent or to tilt back and forth. The block is at the mercy of the water and the wind; its density is modified by the 'choreography of the marine currents and the dance of the air masses'. Here, too, latent heat triggers the melting process. But this accelerates with the iceberg. Nothing could be more normal; the ice is no longer in the 'compressed' state in which it was in the glacier.

The lifespan of an iceberg depends on the ratio of its length to its width, as well as on the vicissitudes of its emerged and submerged parts. The top of the iceberg is initially made of snow or *névé*, while the rest of its body is made of ice. The wind erodes its crest and wears away its exposed sides. Currents and waves crack its sides at and below water level. The whole iceberg is weakened. It slowly rotates on its axis, and the underwater part of the iceberg passes above the waterline. In the vocabulary of Louis Legrand Noble, it is as if the visible 'head' tilts and the 'iceberg of the deep sea' becomes the 'iceberg of the air'. The new peak is all the more prone to erosion by the elements. Then comes the time when it starts to calve itself. It crumbles and breaks into smaller pieces.

Every calving has a 'cost' – that of an irreversible dissolution. The iceberg's days are numbered. The more it drifts, the thinner it becomes. Sometimes it attaches itself to a glacier. It 'reincorporates' and part of it

refreezes. It can also attach itself to a protrusion of the
sea bed. Then it breaks free again and eventually disappears into the ocean. Historian Stephen J. Pyne notes
(in his book *The Ice*, first published in 1986) that, once
in the open ocean, an Antarctic iceberg advances at a
standard rate of 8 to 13 kilometres per day. Its average
lifespan is between four and six years.

Any operation of language produces constraints of
meaning, such that the letter and the spirit do not
diverge too much. Ambivalences are produced, which
must be accommodated. In our case, geophysical fluid
mechanics harbours an animistic tendency: the term
'calving' creates singularity effects. When glaciologists
use it, they assume, voluntarily or not, that the iceberg
is an animate being with its own characteristics. Let us
adopt the latter view for a while and consider the logical
consequences of this way of thinking. There are at least
four.

It is an objective fact that glaciers are ancient.
Contemporary glaciologists agree with John Muir on
this point. Similarly, glaciers periodically release some
of their mass. When they border a sea or a lake, they
calve icebergs at some point in their life cycle. To
describe this phenomenon, scientists add the term 'calving' to the term 'parent glacier'. If the parents are old,
it follows that the icebergs that fall into the water are
also old. They are made up of the same matter. The first
conclusion is that the block that comes off is said to be
'newborn'. But we have to admit that the being starting
to scream is already of a certain age and perhaps even
has the strength of an adult. This reasoning is not fanciful, it is empirical. What kind of newborn could sink a
ship?

A drifting iceberg can be dangerous. Hidden in the
fog, it pierces the hulls of ships and sends them to the
depths. In response to the sinking of the *Titanic* on 15

April 1912, an International Ice Patrol was established
in 1914. Even today, any calving triggers a close monitoring process. This is the case for the tabular icebergs
that break away from the Antarctic ice fronts and follow
the currents of the Southern Ocean flowing from west to
east. Their trajectories are studied by satellite. The relevant authorities issue regular bulletins to the shipping
and tourism industries. Second conclusion: the newborn
iceberg is potentially a killer child, a ruthless predator.

Finally, the iceberg in turn becomes a progenitor.
Calving is not just about being torn away from the
parent glacier. The floating block eventually turns over.
It begins to dissolve. Then it breaks up and leaves fragments behind. Its melting gives birth to several children.
Glaciologists use an informal vocabulary in this respect.
They speak not only of 'parent' and 'child' but also
of 'mother' and 'daughter'. An example: the fragment
B-15B refers to one of the daughters of the B-15 iceberg.
When this happens, the mother iceberg, or parent, is
itself renamed: B-15 becomes B-15A.[30] Thus, the mother
iceberg drifts along with its baby daughters and often
other mothers with their own offspring (small pieces
attract each other). We used to have the habit of seeing
icebergs as loners. Now they become responsible for a
whole family. Third conclusion: the term 'calving' refers
to the changes in the state of the iceberg, from newborn
to adult giving birth to other children. These beings will
then lead their own lives.

It is perhaps easier to see a religious edifice in an
iceberg, as the Romantics did. The function of stained-
glass windows is to give life to the interior of the
human-made structure. One might think that the light
is animated by the air and that the iceberg breathes like
an inspired edifice when the sun's rays slip into its crevices. The imagination works by way of comparison. It
establishes a distance between a subject and an object.
It initiates a hierarchy and draws a dividing line. It uses

metaphors that dissolve the compared into the comparing. The iceberg becomes a cathedral. It is the human artefact that provides the standard for comparison.[31]

It is different when the iceberg and the whale calf are equated, as our little history of the term 'calving' shows. Remember, two non-human entities are brought together. It is always humans who speak. But a relationship of equality is posited between different beings that possess common qualities. No one considers one to be superior to the other. The bodies are not the same. Yet the external attitudes are similar. Icebergs and cetaceans make similar movements in the water. Masses of ice crystals and oxygen bubbles are thus perceived as living in the same way as other beings composed of blood and water.

In his 2007 documentary on the daily lives of people working in Antarctic scientific stations, the director Werner Herzog filmed an interview with Douglas MacAyeal. The glaciologist had been following an iceberg, the so-called B-15, since it broke off the Ross Ice Shelf in March 2000. The proportions of this tabular iceberg were gigantic at the time of the break: 295 kilometres long by 37 wide and 30 metres above the water and 400 metres below. MacAyeal suspects that the block has a consciousness. He knows it well; B-15 occupies all his days, like a work partner. He values it and feels close to it. Herzog makes this moment in their dialogue feel like a liberating confession.[32]

Glaciologists have agreed, at least in language, with the equivalences established by nineteenth-century sailors and fishermen between a material and a marine mammal, as we have said above. It is this equivalence that leads them to describe the world of icebergs on the double model of the ages of life and relationships of kinship. Hence a fourth conclusion: knowing the laws of fracturing glaciers does not prevent us from thinking that the object of study is animate. The feeling that icebergs are endowed with a personality is perfectly compatible with the desire to study them by means of equations and formulas. When scientists watch icebergs calving, they may be thinking about whale calves. Perhaps they are hearing the echo of their own latent animism, like yesterday's sailors off the coast of Greenland.

So, now, only one question remains: does it make any sense?

Do not choose the wrong world

On my return from Iceland, I re-read a text by Val Plumwood which had been haunting my memory. In it, the eco-feminist philosopher recounts her appalling experience with a crocodile. In February 1985, she went canoeing on the East Alligator River in Australia's Northern Territory. She knew that the area was inhabited by colonies of crocodiles. But she ventures in anyway. In the very brief moment before one of the reptiles attacks, the hunter and the hunted lock gazes. She who was paddling along innocently, with no escape tactics or defence plan, had just been spotted by another living creature on the prowl. She is literally pulled under. She survives, despite the violence of the shock and the ensuing hand-to-hand combat. She manages to escape and crawls for hours until rangers pick her up almost unconscious. In retrospect, she interprets her recklessness as thinking she was a superior individual, distinct from other beings.

Becoming another living being's 'feast', turning into 'nutritious food', is not only a cruel and 'very disruptive' experience. When you see yourself in the 'eye of the crocodile', when it takes off, your identity changes radically. You suddenly become the human being as it appears to the predator at that moment, simply prey. The world of wildlife is not 'parallel', it does not unfold

alongside human societies in a universe that is half real
and half imaginary. The only world that exists is the one
where all beings live 'the other death'. Failure to understand this nearly costs Val Plumwood her life.

In her account of this 'initiatory' moment, the author
recalls another episode during which she experienced a
similar feeling of 'disconnection' from reality. During a
canoe trip, this time in the northern reaches of Canada,
she came to a place where the terrain seemed strange.
What she saw was

> landforms . . . strongly marked by parallel strata that
> tilted slightly upwards. Since the human eye is guided
> in these circumstances to take the land as its horizontal
> reference, I experienced a powerful and persistent illu-
> sion that we were moving across a level landscape and
> that the river was running very sharply downhill. But
> there were some things that didn't fit. The river gradient
> seemed very steep, but the water was placid and unhur-
> ried, without rapids.

On the water, Plumwood can no longer grasp the
true connection between the river and its terrain. The
descent may be pleasant enough, but she is worried. She
confuses the cues of the landscape, as if she were moving
in another world. She mixes up the referent of the shoreline and that of the river without managing to relate
either to their real geography. It is not the ground but
the river that is horizontal. The philosopher compares
this peaceful, yet intriguing, expedition to the ordeal of
her encounter with a crocodile. Despite their obvious
difference, both experiences correspond to episodes of
profound disorientation. They are 'moments of truth'.
Each time, she says, the perception of the self goes 'out
of whack'.

Plumwood later developed a critique of the 'hyper-
separation' of humans from non-humans. Despite the
suffering she endured, she learned from her battle with

the crocodile and developed a new philosophy. The premise of this philosophy is clear – to get 'outside the narrative of self, where every sentence can start with an "I"', and finally to see the world 'from the outside'.[33] This hypothesis is applicable to less tragic cases. Beyond the two polar circles, whiteness without a perimeter and expanses without contours shift from a familiar world to one where nothing is understood in the same way any more. Many stories show that living in the ice sometimes brings a feeling of intense presence after the feeling of radical disorientation. The defamiliarisation is such that the experience allows one to clarify some of one's desires and to liberate oneself from the prison of the self. In such places, many have stopped soul-searching. They have forgotten their anxieties and shelved their certainties. Under the influence of living in rough lands, they too saw the world 'from the outside', as if it were new: a world not intact but different, much larger and populated by an infinite number of other presences. They have become part of a planet that belongs to no one.

Events that disrupt one's usual frames of perception must be taken seriously. They change our relationship with the Earth. They force us to bracket out our egocentric viewpoints. They often reveal unexpected affinities between people and unexpected life forms. A 'glacial calving' is of this order. The expression suggests that the glacier is animated and that it is likely to see us watching it. When we use it, we accept a premise of animistic reasoning: to be alive is to be perceived by other beings. The meaning we gain is that of a small revolution in our habits.

Contrary to what we think, we are never alone. Not being seen, not being heard, not being felt – these are impossible erasures. To live is always to be found. There is no doubt a lot more to be done to see, without disturbing them, those beings who observe us, day or night. We do not know how to make ourselves discreet. We hold up too many mirrors between ourselves and nature. Most of the time, we pay attention only to our fellow human beings. Sometimes we celebrate them, sometimes we stigmatise them. Then we run away.

This is the illusion: to believe that we are not scrutinised when we are far from our fellow creatures. To believe, when we are alone, that we move through the world incognito. To believe that we can *really* be alone.

That would be the wrong world.

3

Unexpected Lives

The writer Barry Lopez spent many years in the far north. The result is a finely crafted diamond, a book of wisdom that combines adventure diary with scientific account. In his *Arctic Dreams*, he describes the strangeness of regions with only two full seasons and very high light disparities. The people who live below the Arctic Circle are used to alternating 24-hour periods. Social life follows a circadian rhythm. But, in the Arctic, night does not always succeed day. Much of the year is spent in darkness, a mixture of light and dark that is conducive to sensory illusion. Snow and ice are neither rare accidents nor exceptional situations at these latitudes. They are part of the course of events and of the ecosystem, just as tropical rain or the sand of heat-stricken deserts are elsewhere.[1]

Disordered perception

What does the eye see during polar nights when the mind and body adapt to semi-light? What does the ear hear when the brain is slightly lethargic, a little destabilised by arrhythmic sleep? What do the hands touch when the kind of headache you get on high mountains leaves even a well-prepared individual bedridden?

The great expanses of ice, Northern or Southern, disturb the sensory system.

The monotony of the surfaces distorts the play of distance. You start to see silhouettes. You think you can make out a bear, but it is an Arctic hare. Mirages called *fata morgana* simulate mountainous coastlines. A small sled becomes a huge rocky peak (or *nunatak*). An entire building becomes a snowmobile. Everything seems both far and near. You hesitate. The snow is deceptive.

Differences in scale make you dizzy. Faced with the compression ridges of the pack, the eye is baffled. Rocky outcrops are covered by snow that appears creamy and harmless. What seems flat over there turns out on closer inspection to be a mass of outcrops. Glaciers, with their crevasses, appear behind steep walls. Serac fragments pile up in the hollows. Crystals sparkle above ice bridges and turn into light, sharply defined bubbles in the air. The winds themselves carve sharp ice formations. These are lines of dunes called *sastrugi*.

Sometimes snow on the ground and snow in the sky mix. Sometimes clouds are traversed by rainbows. Sometimes the panorama becomes uniformly white. The 'whiteout' phenomenon is familiar to mountain people and is also common on sea ice. The reflected light is reduced, landscapes become absent and shadows fade away. The background disappears as it merges with its opposite. As horizontal and vertical lines coincide, one is blinded by a totally flat clarity. The sense that usually dominates elsewhere, the eye, admits its powerlessness. The other senses waver. In the mountains, you have to look at your feet and wait for the summits, high up, to reveal themselves in order to secure landmarks. Those who travel on sea ice must have an attentive ear in order to hear, for example, the sound of waves without seeing any coastline.

Often behind the ice, there is yet more ice. A whole expanse before you and extending as far as the eye can see. You can see it without understanding it. During his Antarctic expedition of 1901–4, the explorer Robert Falcon Scott confessed that 'it is not what we see that inspires awe, but the knowledge of what lies beyond our view. We see only a few miles of ruffled snow bounded by a vague wavy horizon but we know that beyond that horizon are hundreds and even thousands of miles which can offer no change to the weary eye.'[2]

Ice disorients. It imposes itself by abolishing most of the usual responses and indicates a temporary failure of the analytical mind. Familiar ways of apprehending time and space are neutralised. Think of the aurora borealis or the aurora australis, with their moving theatre of huge coloured curtains that reduce us to silence.

Frozen surfaces seem so plain that adventurers sometimes do not even consider them to be offering any features at all. Carl Weyprecht, the sea commander of the Austro-Hungarian expedition described by the writer Christoph Ransmayr, wrote in his diary:

> Whoever will truly admire nature should observe her in her extremes . . . In the tropics, the eye loses itself in the multiplicity of details to be admired. But here lack of detail causes the eye to be directed at the overwhelming whole, at her productive powers in the absence of things produced. Here one's attention, freed of preoccupation and unmoved by particulars, can be concentrated on nature's powers.[3]

Weyprecht speaks the language of the sublime and elemental powers. He believes that the monumental nature of the North does not bother with signs. Instead of being lost in the order of the minuscule, it unfolds the terrifying order of the immense. In the middle of the ice, no one can expect that an opulent setting, rich in clues, will emerge to catch their attention. Everyone is overwhelmed by the material excess. At the time the lieutenant of the line is reporting these observations, the crew is in a crisis situation. They are trapped in the ice. The universe that stretches out to infinity reflects their deep despair.

For them, nothing lives. Death rules.

Whatever the experience, the worlds of ice and snow disturb our certainties. They decivilise the gaze. No one can cheat their convictions, or even their senses. It is a test of truth. When the bottom and the surface are no longer distinguishable, an alternative presents itself: inner contemplation or a body-to-body encounter with the ice and the sky, a filtered portrait of light.

Crystals fly through the air on a clear day. Then they come together and form small clouds. They diffract the light through iridescent crowns. They dance and have a slightly pungent, almost salty smell. Perhaps this is why the Chukchi of eastern Siberia welcome the first snowflakes that fall.[4]

Others believe that the glacier is looking at humans, the sea ice is waking up and the iceberg is screaming like a newborn baby. In the land of snow and ice there is a push to recalibrate the boundaries between life and death. The waterline and the submerged sides of a berg drifting in the ocean are worth observing carefully. Ecological processes that are constantly being renewed cling to them, contradicted by the impression of a frozen equilibrium. Since everything is white, everything seems static. However, lives are suspended in the basement of the iceberg, its submerged part. They bubble up to the surface of the water and attract marine mammals.

Living ice

Norway, in the year 1250.

A son asks his father about the nature of the ice and icebergs in Greenland and Iceland. This is not his first question. He wants to know more about these distant countries, two almost uninhabited islands, battered by the waves. He urges his father to inform him with the utmost precision. As if planning an adventurous journey, he wants to know about ways of life, the wind patterns, the strength of the currents, the conditions of agriculture and animal husbandry. The frozen masses on land and at sea intrigue him. The father regards his son for a long moment, then replies.

He tells him that these blocks have a 'peculiar' nature. Sometimes they are motionless, sheltered by great fjords, sometimes they move like agile animals braving the winds and gliding across the ocean. Some ice has very particular shapes. The Greenlanders call them 'fallen glaciers'. Icebergs look like 'mountains rising out of the sea'. They 'never mingle with other ice, but stand by themselves.'[5] With his finger raised, as if to arouse the child's curiosity, the father adds that whales and seals follow in the path of icebergs. The marine mammals probably find plenty of food there.

This educational dialogue written in Old Norse is addressed to King Magnus VI of Norway, Lagabøte. The drifting blocks are described as solitary beings. But the father guesses that they carry life forms that other animals covet. The hypothesis of a rich ecosystem has since been widely verified.

Every ice shelf is an 'ecotone', Barry Lopez and the biologists tell us, a 'transitional area', a subtle medium of exchange between biotic communities. When dust falls on an ice cap from the lower atmosphere (the troposphere up to 15 kilometres and the stratosphere up to about 50 kilometres above the earth's surface), it absorbs heat. The heat creates cavities in the ice that act as a multiplier of life, as in an aquarium. Ice algae cling to these 'biological folds'. Tiny diatoms, and all the bacteria that colour the sides of the glaciers, are the zooplankton's favourite food. These are then eaten by cod. In the Arctic, the cod in turn become the prey of

northern fulmars, narwhals and seals. Seals are hunted
by polar bears. The polynyas, these 'areas of persistent
open water', represent a kind of Eden in the sea ice for
phytoplankton to proliferate. Guillemots, eider ducks,
beluga whales and Greenland whales rush to find food.
In short, ice is a building block of the great food chain.
It is to ice that all 'this life owes its cohesion'.[6]

The process is repeated on the scale of the iceberg.
This is seen with each rotation of the iceberg. When
the submerged part of the iceberg tilts, the sea begins to
rustle. Air escapes. Fresh water mixes with salt water.
Phytoplankton are released, which feed the zooplankton. And the whole cycle starts again. As the icebergs
break off, they also reveal seabeds that have long been
preserved under the thick ice. Soils are rich in microbes.
They are home to other chemotrophic organisms, which
thrive without sunlight, as well as to molluscs. Whatever
their size, icebergs not only support the food chain, they
help organise the life process itself.

These remarks apply to both the Arctic and the
Antarctic. At first glance, however, the White Continent
looks like a 'desert landscape'. Even those familiar with
snow can still feel this way. In one of his adventure
stories, the explorer Wally Herbert reports the genuine
astonishment of a group of Greenland Inuit who were
shown a film about Antarctica. They are stunned and
huddle together as if they are afraid or can no longer
tolerate the freezing blizzard. Wide-eyed, they see 'only
a cold desert, beautiful but barren. There was no vegetation there; no gnats, mosquitoes, mice or hares; no musk
oxen, reindeer, caribou or polar bears. It was a weird
world they had seen in these pictures, desolate and pure
– quite unlike their living, breathing, hunting territory.'[7]

The status of Antarctica, as we know, is unique. No
indigenous population has ever lived there. Religious
missions, armies of mercenaries and curious anthropologists have ignored this part of the world. Its myths

were first brought back by explorers, then by scientists holed up in ice-whipped stations. Signed in 1991, the Protocol on Environmental Protection to the Antarctic Treaty applies to land and ice areas south of 60 degrees south latitude. It makes the frozen continent a vast nature reserve.

Compared to the Arctic, the biological environment of the Antarctic looks to be less rich; it may well seem desert-like, even to the Inuit. But all deserts are full of unexpected life. Krill is exceptionally abundant. Adelie penguins and skuas, predatory birds, inhabit the pack all year round. Leopard seals occupy the coasts and islands. Whales flock during the southern summer. To interpret their way of life, we must be more patient than ever, educate our eyes and learn to change focus, to move from the panoramic to the imperceptible, from the colossal to the infinitesimal. The eye must anticipate sudden changes in perspective in order to orientate itself in the distance as well as in the near future. In Antarctica, the katabatic winds, the air currents that sweep down the slopes along the coast, are cold and powerful. They blur the view and turn the horizon upside down. In such places, as we have said, our various responses acquired during peregrinations in temperate zones are inoperative. Another perceptive device is required. Scuba diving remains an effective practice, revealing invisible worlds.

Barry Lopez dived with scientists specialising in the study of organisms that thrive at depth – so-called benthic ecology. In his last book, he recounts the moment when he walks along the sides of icebergs whose 'keel' has 'stuck' to the seabed (grounded icebergs). The group is swimming in a bay not far from McMurdo Sound in Antarctica. All of them immediately notice the contrast between the naked icebergs above the waterline and the teeming organisms just below. In the first few metres, an entire community of organisms emerges: starfish, sea urchins, scallops, nemertes (or marine worms), and sponge algae that hang from the submerged walls.

By moving under the sea ice, divers discover an incredibly rich living world that the currents constantly carry, move and replace. They know that the passage of icebergs scraping the ocean floor redefines the state of the benthic environments. The large masses plough up entire areas, taking part of the soil with them and forcing species to adapt to their movements. Everything deforms and re-forms at the same time. Everything breathes together. Lopez understands that the ice acts as a 'keystone species', in the vocabulary of biologists. It carries and spreads life around it via a chain of reciprocities.[8]

On the surface, too, everything moves and nothing is still. Snow and ice provide shelter for ermine and especially lemmings (which would not otherwise survive). By keeping plants warm, they protect them from the wind, which suffocates and torments them. They allow seals and walruses to move around to feed, mate and care for their young. They serve as a 'winter migration route' for musk oxen, caribou, bears and polar foxes.[9] The seasonal cycles of ice and animal and human migration are closely intertwined.

Ice shelves, glaciers, icebergs, all these masses are dynamic and interdependent structures. They amalgamate various types of snow and ice – sea ice, land ice, surface snow. They adjust to shorelines, submarine landforms and their life forms. The essayist and writer Gretel Ehrlich emphasises that the arctic ecosystem 'co-evolves with the ice'. Ultimately, sea ice behaves like the 'Amazon canopy'. It similarly provides 'a ceiling and a shelter'. It 'gives nourishment' and 'creates its own weather'. In other words, ice is a 'home' for most of the beings that live there. One of the differences with the Amazon, however, in addition to the number of species (unmatched in the rainforest), is that the fragility of the ecosystem is 'hard to see, since so much of the life occurs under the ice.'[10]

Iceberg portraits

Barry Lopez has travelled the lands of the far Northern Arctic, sailed along its coasts, met its indigenous people, and even dived in Antarctic waters. He describes icebergs like a painter. The bergs become cubes of colour under his pen. They alternate translucent white with navy blue, when the ice is old, or a slightly greyish blue-green, when it is more recent and already cracked by wind and surf. The low-slung sun adorns the icebergs in large purple sails. The colours change according to the angles of its rays, sometimes oblique, sometimes vertical, and the state of the ice crystals that reflect them. If clouds gather, the shadows reinforce the contrasts and create a subdued atmosphere that blurs the shapes. Around icebergs, the water is often opaque. The descriptions could be multiplied to remind us of the extraordinary capacity the bergs have to create a luminous geometry. The range of colours is infinite, open to all prismatic variations. This is what seduced the romantic travellers.

But is light enough? Lopez also has something of the exploratory naturalist in his character. He travels to enrich his knowledge of the affinities among living beings. He solves the enigma of icebergs that were thought to be dead: the drifting blocks are biological centres. Better still, they produce life. And this life unfolds between the play of light on the surface and the darkness of the depths. The writer knows this. He has seen it and told the story. Icebergs are the driving force behind his passion for the Arctic. He is filled with happiness when he sails near them, seeing and hearing them.[11]

An exclusive focus on light would lead one to believe that icebergs pit the ocean against the sky. This would be a mistake: they constantly mix the two elements. This would also be to forget that icebergs carry biotic communities, to which some human beings feel they belong. Their dramatic properties, ever-changing shapes, bird calls, wave colours and other signs point to a wider and older kinship.

In a 2008 exhibition at the National Academy of Sciences in Washington, DC, Camille Seaman presented images of icebergs taken during her travels in Greenland and Antarctica. Since *The Last Iceberg*, she has had the opportunity to take other shots. All her photographs are portraits that capture the personality of the frozen masses. In a TED talk on 16 June 2011, she explains her approach.[12] She says:

> It's easy to think, when you look at an iceberg, that they're isolated, that they're separate and alone, much like we as humans sometimes view ourselves. But the reality is far from it. As an iceberg melts, I am breathing in its ancient atmosphere. As the iceberg melts, it is releasing mineral-rich fresh water that nourishes many forms of life. I approach photographing these icebergs as if I'm making portraits of my ancestors, knowing that in these individual moments they exist in that way and will never exist that way again. It is not a death when they melt; it is not an end, but a continuation of their path through the cycle of life.

After commenting on her own photographs, the artist shows the audience a short film sequence of an iceberg, approximately 40 metres high, turning over in a few seconds.

It looks like a whale rolling over.

No being is opposed to another being for those who know how to see and listen, nor any element to another element. Camille Seaman defines herself as a 'citizen of the Earth'. She is attached to the land where she was born. Her father is from the Shinnecock tribe (who

live on the tip of Long Island in New York state). Her
mother is of African-American descent. Both her parents always told her that nature and culture were not
separate and that humans had no privileges over other
living things. She has made this her philosophy. For this
polar photographer (who has since taken an interest in
tornadoes), icebergs have their own stories and characters. Their wanderings are punctuated with moments of
drama. Some 'refuse to give up and hold on to the bitter
end, while others can't take it any more and crumble in
a fit of dramatic passion.'

In the three series of *The Last Iceberg*, Camille Seaman
photographs the persistence of icebergs, right up to their
final spasm. The images are not only beautiful, they are
also profound. Through their grain, their clarity, their
angles, they reveal the fragility of the enormous blocks
that are tilting and preparing to fall. They take us into
the crevices of the glaciers. They almost allow us to
anticipate the sound of their breaking up. We discover
former submerged parts of icebergs that have turned
over. Soft lines appear, all creamy, volutes polished by
underwater currents, like the traces of an ancient caress.
The walls are also festooned with cells carved by the salt
water. Several images are aerial. We are sitting on top of
tabular icebergs. We pass between two giant pillars with
dull, opaque colours, like snow petrels. The photographer pushes us in the back so that we can quickly get
some air before gigantic slices of ice fall into the ocean.
This time we can clearly hear their creaking. Behind us,
coastal ice fronts contain cave entrances that seem to
run to the centre of the earth.

Such images capture the intimacy of icebergs. They
make them seem closer, even when they stand at a
distance. The otherness that emanates from their out-
landish proportions seems much less radical. It becomes
confidential, like a whispered word. How can one
describe the feeling of having a nature in common with

Figure 3 Camille Seaman, *Stranded Iceberg*, Cape Bird, Antarctica, 2006.
© Camile Seaman Photography

Figure 4 Last Iceberg, Series I, II, III
www.camilleseaman.com

an iceberg? The eye is not enough on its own. Camille
Seaman knows this. That is why her images make us
hear voices. They make different tonalities resonate.
The sounds are sometimes harmonious, sometimes dis-
cordant. They suggest a symbiotic experience while
preserving a gap.

Icebergs are snowflakes piled on top of each other,
compounds of water. Our bodies are 70 per cent water.
The photographer believes that icebergs are humans,
perhaps even distant relatives. For her, too, any break
in a glacier is a birth. But it is the birth of her ancestors.
We can guess at the hypothesis at work here. Whatever
beings we are talking about, the law of life is identical;
the generations renew themselves and mix. Like us,
icebergs are born and die. They react to what affects
them. We share the same fate as them. We share breathing (the same air), vital circulation (water as a universal
medium), the kinship among species (ancestral icebergs)
and the experience of disappearance (melting). All these
aspects form, when put together, a kind of intense, wild
magic square, uniting humans and non-humans.

Seaman's portraits are visual examples of sentient
icebergs: the bergs communicate the palpitations of the
beings around them, like yet another breath found, or
found again, beyond any effort. If one of her ancestors
were to appear on images in a videographer's montage,
it would not be in the form of a skull. Nor would it
be a reflection of herself. It would be a face with lips

articulating a soft melody. The text might say that icebergs are taking their place in the long genealogy of relationships among beings, that this multispecies filiation cancels out the myth of their solitude and that there is no mirror between them and us.

Looking at the photographs in the *Last Iceberg* exhibition, I thought I could hear the floating masses breathing. I wasn't freezing cold, alone in the middle of the ocean, but lulled by warm voices.

All power to the verbs!

Camille Seaman's photographs are synaesthetic. They solicit and unite all the senses. The power of the images should not push us to cut short our thinking on language. Many people also try to translate relationships of a broader scope with whatever words they can find. The task is never a given. Using words to describe a natural phenomenon involves linking elements together. We always run the risk of fabricating in language relationships that do not exist in nature: which is cause; which is effect? There is a danger of insinuating hierarchies where there are only solidarities.

During his Swiss trip in 1775, Goethe wanted to get around this difficulty. He juxtaposed words without naming the relationship in order to depict the imposing mass of the Saint-Gothard: 'Snow bare rock and moss and tempestuous wind and clouds the sound of the waterfall the tinkling of the mule.'[13] He made punctuation and verbs disappear, leaving it up to the nouns to capture the essence of things. Would such noun phrases allow the life of an iceberg to be told as some images do? That would be wishful thinking, a passing fancy.

To speak of life, verbs are more useful than nouns.

In a humorous text, academic and writer Robin Wall Kimmerer tells us how she learnt the Potawatomi language, spoken by her ancestors. She keeps nervously flipping through the dictionary. She doesn't understand anything she reads. Too many things are like verbal states: 'to be a hill', 'to be red', 'to be a long', 'to be a long strip of sandy beach'. She wanders through the corridors of incomprehension, stumbling across so much confusion.

One term, however, catches her attention. It is the word *wiikwegamaa*, which means 'to be a bay'. She smiles, guessing something, but the language seems so 'heavy', 'complicated' and 'impossible to learn' that she thinks it's all too hard. She even begins to think that this language is 'false'. Bay is a noun! But the term that describes it refers to a verbal state. It's enough to make her give up the game. She imagines the ghosts of former boarding school missionaries rubbing their hands in contentment. Yet another triumph of English over a vernacular language!

Suddenly, she feels a revelation flash through her body at the speed of light. She comes to her senses and realises that she is breathing in the smell of the water filling the bay. She sees it hit the shore. She hears it lapping at the sand and pebbles. A magical moment. From this unique perceptual experience, she draws a general reflection:

A bay is a noun only if water is *dead*. When *bay* is a noun, it is defined by humans, trapped between its shores and contained by the word. But the verb *wiikwegamaa* – to *be* a bay – releases the water from bondage and lets it live. 'To be a bay' holds the wonder that, for this moment, the living water has decided to shelter itself between these shores, conversing with cedar roots and a flock of baby mergansers. Because it could do otherwise – become a stream or an ocean or a waterfall, and there are verbs for that, too. To be a hill, to be a sandy beach, to be a Saturday all are possible verbs in a world where everything is alive. Water, land, and even a day, the language a mirror for seeing the animacy of the world, the life that pulses through all things, through pines and nuthatches and mushrooms. *This* is the language I hear in the woods; this is the language that lets us speak of what wells up all around us.[14]

The missionaries are not happy. This time they have lost. They think with names instead of living with verbs. Today, the Potawatomi language is spoken in the United States (Michigan, Kansas, Wisconsin) and Canada (Ontario). It is composed mostly of verbs (70 per cent compared to about 30 per cent for English, according to the author). It does not really distinguish between what is masculine and what is feminine, unlike other languages, which systematically separate genders. Verbs and nouns are both animate and inanimate. Verbal states have the value of personification or, rather, incarnation. The neuter plays a minor role. In the end, it is context that counts the most.

What Kimmerer calls a 'grammar of animacy' is the set of principles and rules that order the system of vocal, graphic and conventional signs used by her ancestors. In order to explain what she means by this expression, she draws examples from everyday life. It is not conceivable to say of the grandmother who prepares the soup in the kitchen: 'Look, it's making soup, it's got grey hair.' This would be a way of stripping her of her personality and undermining her dignity by turning her into a thing. The 'it' in this case is a linguistic trait: the 'it' takes away life. Designating living things with pronouns that make them inanimate is a clear lack of respect.

The Potawatomi language, Kimmerer tells us, uses 'the same words to address the living world as we use for our family.' The reason for this is simple: it is because the majority of beings on the planet 'are our family'.[15] It follows that words are not just used to designate realities. They translate the life choices beings are making; water, for example, which prefers the society of large creeks to the emptying out of rivers. Above all, they reveal relationships of filiation.

In the Potawatomi language, kinship seems to extend to all the dimensions of the Earth. What are its exact limits? Plants, animals are naturally part of it. Rocks, mountains, water, fire and places are also alive. 'Beings that are imbued with spirit, our sacred medicines, our songs, drums, and even stories, are all animate.' In contrast, the list of inanimate objects is short. It includes artefacts and objects made by human hands, such as a table or a chair. The maple tree is a living being. This should be remembered before tapping it for firewood: 'Saying *it* makes a living land into "natural resources". If a maple is an *it*, we can take up the chain saw. If a maple is a *her*, we think twice.'[16]

So how do we appreciate all beings at their true value? Perhaps the first rule is this: turn to our ways of speaking, flush out the logics of commodification that nestle in our everyday words and neutralise the neuter. Kimmerer believes that languages that value animacy allow us to remain attentive to the world around us.

In general, languages induce behaviour towards other beings. The ecologist David Abram states that they are a 'means [not only] of attunement between persons, but also between ourselves and the animate landscape.'[17] Kimmerer's ancestors would certainly agree. Let us apply the argument to icebergs and glaciers: they are not things but living entities in the cosmologies and languages of the Arctic peoples. To regard them as mere sources of fresh water would once again fail to accord them the dignity they deserve.

Incorporation/orientation

It is not essential to know, as you may have guessed, whether icebergs and glaciers can be considered on a strictly scientific level as living beings, nor is it to advocate the virtues of animism as an integral system of representations. In the first case, the answer would undoubtedly be negative. In the second case, it would consist, for many, of taking up a kind of attitude. The main point is to show that the roles of glaciers and icebergs are multiple and that the boundaries between approaches are porous. It is a matter of identifying the characteristics on which everyone can ultimately agree. To do this, it is necessary to develop interpretative frameworks that give the senses their full rights again.

One impression dominates as you walk across the flat expanses of the far North. The silhouette of the trees shrinks as you go up in latitude. Vegetation retreats. The number of species is reduced. The uneven distribution of light makes photosynthesis more difficult. Cold winds force living things to move closer to the land, where temperatures are milder. Since water is scarce, resistance tactics are essential. Barry Lopez tells us that Richardson's willow and the dwarf birch are among the only trees that can withstand the ordeal of permafrost, the solid, impermeable frozen ground. You have to kneel in the peat and look at the soil that feeds the glaciers by capillary action. You can see that there are trees that live flush with the grass and grow horizontally. They are no taller than the fingers of a raised hand and can live as long as a maple tree in temperate zones, about two hundred years. The number of their growth rings testifies to this. To walk on the tundra is to understand that you are 'wandering around on *top* of a forest.'[18]

Contrary to appearances, the soils of the far North have a biodiversity of their own. Plants change colour in summer to absorb a wider spectrum of light. Arctic bumblebees are very active pollinators. The tundra is covered with an infinite number of small clumps: mosses, sedges, saxifrages. The inattentive eye will not notice any forests in these parts. They shrink so much in order to survive that they turn us into Gulliver in Lilliput. We become giants. But the Arctic is not the product of a writer's inventive fancy. It is not an imaginary island but a collection of very real lands. These lands force us to conjugate differently the relationship between the visible and the invisible, to perceive the passing of time differently. Locating the forests that lie beneath our feet and that we cannot see with the naked eye requires us to adjust to the particular rhythms of this environment. You have to be able to wrap yourself in the garments of wind and snow.

In 1888, the anthropologist Franz Boas provided a synthetic view of the representations and myths of the Inuit population. In a passage in his book *The Central Eskimo*, he tells the story of a birth. What does the mother do with her newborn child? She wraps it immediately in the softest fabrics. She sews underwear with the first feathers of fledglings and uses the furs of young hares to cover its head. She inserts it into the 'country' from the very first moments. It is crucial that the child senses the smell of other beings on its own skin at an early age and that it also hears the pack, the thick sea ice that squeaks and rubs with sounds similar to 'the whining of puppies and the swarming of bees'.[19] The baby becomes intimate with its environment.

The elemental landscape is inscribed in the bodies of the Inuit. The relationship to a place is first and foremost physical. Far from disappearing with time, it develops over the years as an inalienable presence in the flesh. The sea ice can take any kind of bad weather or catastrophe. At least, that's what used to happen a long time ago. In recent decades, it has been shrinking. If it were to disappear completely, the world itself would collapse. For it is part of an even larger environment. And, in the eyes of the Inuit, this environment is *a priori* eternal. Basically, explains Barry Lopez, the frozen landscape fulfils the same function as architecture for us; it situates the spirit in time and space. It gives a sense of duration and allows the indigenous people to hope that the universe itself will never end.

This natural setting carries human stories. The example of architecture does not play the same role here as in Louis Legrand Noble's travelogue. It does not offer a whole range of postcard-like monuments. It makes it clear that the ice, however bare it may seem, is an endlessly legible territory. It contains the memory of events that took place there. To walk on it is to reactivate the deep memory of the links that humans have with each other and with the Earth. At these latitudes, all places are memory places. Even in a hospital room, when one is alone. An Inuit woman tells her interviewer that she places her hands in front of her eyes and a whole book of vivid images opens up. She sees in her mind the mountain passes, the shores and the coves of the sea that she has visited, and hears the calls of the birds and the growls of the seal. Despite her destitution and loneliness, this patient retains the strength that her environment has given her. She is separated from her native soil, perhaps forever, but her memories of it are intense, carnal as well as spiritual. Like compass bearings.[20]

Lopez is interested in the strategies that the Inuit use to move forward in a space that seems to him to be devoid of any landmarks. He notes that the hunters do not pay much attention to distances and that they are concerned more with the small details they identify in the landscape and that other hunters have sometimes mentioned before their departure. Accompanying them on the ice pack is a bit like becoming an 'Arctic fox': you constantly look back, you often retrace your steps, there are long pauses (to warm up, drink or eat, and feel the wind), and there is never any question of moving forward in one go towards a specific 'goal'. Those who are not used to it will see a lot of wandering, maybe even a strange dance. But it is another way of finding your way in an environment where one of the major risks is getting lost.

The writer shows that certain climatic conditions force one to find one's way with a minimum of signs and that one must therefore pay attention to the smallest signs. Even in an entirely white landscape, when the contrasts fade or disappear, traces still exist. Even in the middle of a blizzard, day or night, there are still footprints. You can't see them, but they are there. Nothing is completely erased. The Inuit hunter, moving forward in a complete whiteout or in the dark, listens for bird calls, observes the fur on his clothes and strokes the snow ridges on the ground in order to interpret the changes in wind direction. Any details that might relate to landmarks already noted in his mental maps are useful. The task is endless. Every clue must be noted, certainly not confused with any other, and kept in mind. A simple scent can be used to find your way back. Time is so versatile that everything counts, nothing is superfluous. It is a question of survival.[21]

Orientation strategies must be effective on the ice. Those unfamiliar with the pack ice will tend to orient themselves as if in a city. They will look for obvious features, points that catch the eye. They will probably get lost. Anthropologist Edmund Carpenter's first impression of an outing on the ice is that his fellow Aiviliks (Inuit from Canada) develop reflexes roughly equivalent to his own. They just focus a little more on natural elements. Carpenter means that everyone seems to trust what they see. But this is not the case.

The edge of the horizon blurs and visibility is reduced. His friends manage without their eyes. They move without recourse to objects or signs that the eye can recognise. They use their knowledge of the relationships between 'contour, type of snow, wind, salt air, ice crack'. The anthropologist (to whom Lopez himself pays tribute) goes on to give an eloquent example: 'Two hunters casually followed a trail which I simply could not see, even when I bent close to scrutinize it; they did not kneel to examine it, but stood back, examining it at a distance.'[22] The hunters in question do not bend down to look at the ground but straighten up, the better to make out the extremely faint outline of an old line in the snow. They find a perspective where there no longer is one. However, it is not the case that the indigenous people can see any better than other people. Their eyes are not superior. Their way of seeing the ground is a training passed down from generation to generation. It means something to them because they have acquired the know-how. Carpenter realises that he has always been an 'observer', as if trapped in a double-locked ivory tower. They, on the other hand, 'participated' in each of their expeditions on the sea ice with all their senses.

Barry Lopez tells a similar story, an old memory. He accompanied some indigenous friends on a trip into the taiga. One day they come across a grizzly bear sucking on the last remains of a carcass. This is an opportunity to compare ways of interpreting the event. He focuses on the animal, its movements, its size, the colour of its fur. He immediately tries to translate the experience of this encounter into a syntactic framework. He wants to make sentences to fix the meaning, a normal reflex of the analytical mind.

His companions have a different, more synthetic attitude. They recall seeing caribou hoofprints on the ground along the way and also noted unusual variations in birdsong. They heard the crackling of brush and saw tufts of hair clinging to the bark of trees. They are reactivating their sensual memory and placing the episode in a wider context. The wild animal becomes a 'fragment' in a process that began before this encounter between humans and non-humans and will continue afterwards. No meaning is fixed. Lopez's friends see themselves as part of the same sequence of events as the bear. Whereas, for him, 'the bear was a noun, the subject of a sentence; for them, it was a verb, the gerund: *bearing*.'[23]

The episode is significant. One of the virtues of the ethnographic gaze is to learn to relativise the categories we use to shape the environments we inhabit. This is what Lopez does. He grasps the differences in the strategies employed. His are riveted to the order of words; he has to write, or speak, to define the event and circumscribe it within a geographical perimeter. The writer postulates that the world reveals itself when put into language. For others, on the contrary, it does not need to be represented. In the words of the anthropologist Philippe Descola, Lopez's and Carpenter's companions establish 'relations of correspondence and opposition between salient features' of their environment. These traits are qualities of the environment that they 'actualise' by responding to the 'ontological choices' that seem most relevant to them.[24] Their choices are part of a system of thought that inscribes the human in the non-human.

One suspects that the boundaries between the living and the non-living vary between cultures and groups of people. Sometimes this opposition is blurred or simply doesn't exist. Tim Ingold reminds us that, for many indigenous people, 'life is not an attribute of things at all. That is to say, it does not emanate from a world that already exists, populated by objects-as-such, but is rather immanent in the very process of that world's continual generation or coming-into-being.' The anthropologist rejects definitions that make animism a set of beliefs diffusing life into things. Life is not a predicate. It is not granted, in the way a speaker attributes a property to an object in assertive statements. One does not perceive the world by projecting imaginary properties onto objects that the mind has constituted. Animacy is not 'the result of an infusion of spirit into substance, or of agency into materiality, but is rather ontologically prior to their differentiation.' In short: life is not a 'way of believing *about* the world', which language would make explicit; it pre-exists any distinction between subject and object.

Ingold criticises the concept of surface. This notion implies that one moves on the ground of the world as on a map composed of points. It is better to say that we walk through places following trajectories in changing volumes. Arctic lifestyles confirm this approach. On the ice, we move along lines that make us evolve in an atmosphere where the elements are constantly mixing. You can't find your bearings in a confined space. Most often, there is no clear circumference, no distinct outline. The Inuit say that 'the trails they leave behind' allow people to be recognised. Everyone leaves a multitude of signs and clues in their wake.[25]

Sculptures made in the Arctic express this intensely participatory life, always in dialogue with the environment. Carpenter is especially fond of artefacts representing seals. Such objects show that the craftsman 'becomes one with the seal, and thus succeeds in portraying it, for he is then a seal himself.' In the same way, his companions succeed without difficulty in 'parodying' the beings that surround them: 'bear, iceberg, yes, even wind!'[26]

When it comes to icebergs, we can already imagine several parodies: an imitation of how upset it is when it melts too quickly, how it could, like a whale calf, be dismayed if it fails to control a roll in the water, or even a fall into the water that would make it spring back with

such great force that it would give the impression of
flying up into the sky.

A world that was thought to be immobile comes to
life. This simple game implies a different understanding
of the relationship between the visible and the invisible.

Noise and breath

At first glance, an iceberg panorama might look inert.
But when a glacier calves or a floating mass tips over,
it is no longer a static scene. In birth, or rolling over,
icebergs can be very expressive. They are a reminder of
how intense ice is as a material, both hard and vibratory.
This kind of event makes it impossible for the viewer to
represent exactly what is happening. The volumes are so
powerful that they connect us, brusquely, to a strange
world. They also keep us at a distance. Several minutes
go by; the words are still missing, as if the vocabulary
had lost its usual roots. Where are the figures of speech
to describe the sudden animation of a body to which no
life is usually attributed? Everyone comes to realise that
language is not enough. The only way to get the upper
hand is to pay attention to the sounds.

We already know through John Muir that the occa-
sions of icebergs being born are great moments of social
acoustics. The Tlingit guides who accompanied him
on his Alaskan expeditions commented on the iceberg
break-up events they witnessed together. The terms they
use give new life to the icebergs. This is how Muir reports
their exclamations: 'And while berg after berg was being
born with thundering uproar, Tyeen said, "Your friend
has *klosh tumtum* (good heart). Hear! Like the other
big-hearted one he is firing his guns in your honor."' Or:
'When the Indians came ashore in the morning and saw
the condition of my tent they laughed heartily and said,
"Your friend [meaning the big glacier] sent you a good

word last night, and his servant knocked at your tent and said, '*Sagh-a-ya*, are you sleeping well?'"[27]

The explorer-naturalist shares the observations of his Indian guides. The roars of the glaciers and icebergs are like words that listeners, like him, of course, as well as others, can understand. You have to listen to them and answer their questions. The masses speak to an audience of humans and non-humans alike. The sounds they emit make up a varied vocabulary that travels from one glacial valley to another. They introduce vocality into the silence of the great outdoors. In their outbursts, they tune nature, like musicians in an orchestra. They enliven it with vocal bursts that carry archaic echoes, perhaps the oldest sounds ever heard. With icebergs, the world is not produced by human intelligence, it plays its own music. A bay becomes a 'soundscape'.[28]

Glaciers greet their audience with iceberg melodies, sending friendly signs to their hosts. However noisy and intimidating, the birth of a block of ice is a declaration of hospitality, a permission to remain in the territory, like a special ticket to enter a restricted area. As it begins its symphony, the orchestra of giants begs the guests to enter into the conversation. Glaciers have a visible sound signature; the cries of icebergs falling into the water are their 'visual voices'.[29]

Muir and his guides know that the ice is always moving. It never sleeps. Perhaps it even has expressive dreams in its sleep. The world of glaciers and icebergs is not silent. On sea ice, if you have a good ear, you can distinctly hear the sound of hammers, sometimes birdsong, or the clanking of chains. All these sounds are due to the pressure of the ice.

The blocks produce other sounds that humans can hear only if they pass below the waterline. Only hydrophones placed in shallow waters, with the help of satellite systems, can capture the variations of these acoustics. Icebergs have various sound lines, leaving individualised traces. Oceanographers take on the job of tracking and recording them.

Most of the studies focus on icebergs drifting from fragile sites in Antarctica. They are biggest there, and their trajectories are more significant than elsewhere. Scientists are picking up sounds that propagate in a manner not unlike earthquakes. The South Pacific is filled with the sounds of erratic icebergs. Modulations that start in Antarctica can even be heard as far away as the equator.

Acoustic oceanographers tell us that two types of sound are generally identifiable: long harmonic vibrations, on the one hand, and, on the other, short bursts indicated on the spectrograms by less dense frequency bands. In the first case, the sounds result from collisions with other icebergs, rubbing against underwater protuberances, or even momentary anchoring in the shallows. In the second case, they indicate rapid disintegration phenomena, internal fracturing in the open sea, or intense erosion of the edges.[30]

Hydrophones perform a 'stethoscope'-type function. They are used to listen to the rhythms of an aquatic environment, to take the pulse of a drifting mass and to anticipate the slightest breakages. They are also used to measure the speed of wave penetration, to assess the rate at which an iceberg is melting and, depending on the case, to diagnose its impact on marine animal life. Physicist Philippe Blondel says it's like working with a 'sound mixer'. Like a sound technician with a set of string and wind instruments, he first identifies the sounds made by icebergs. He registers their timbre, intensity, frequency and duration. Then he links all the iceberg sounds to establish a vocal identity. In the end, he obtains a wide range of sonorities that alternately evoke sparkling water whose bubbles disperse in the air, the song of whales, a house cracking in the sun or the drone of an aeroplane.[31]

Petra Bachmaier and Sean Gallero (otherwise known as Luftwerk) wanted to bring the amazing sound trajectories of icebergs to a wide audience. Working closely with Douglas MacAyeal, they produced an installation in Chicago early in the autumn of 2017. The glaciologist, already filmed in the aforementioned documentary by Werner Herzog, collected an impressive amount of seismic data in Antarctica between 2001 and 2007. The two artists placed four watertight baffles on the North Riverside Plaza. The speakers played a tape loop of calving and block icebergs drifting in the Southern and Pacific Oceans. One of them is particularly eloquent. It is the one recording a birth from the Larsen C ice shelf. On 12 July 2017, a fracture with a line almost 200 kilometres long released an iceberg classified as 'A-68'. The behemoth, renamed 'White Wanderer' by Luftwerk, then began a voyage that is now complete.[32]

The artists wanted to familiarise the non-specialist public with issues related to the climate crisis. They created an environment of 'interactions' between different sound worlds: the concrete city and frozen masses roaming the ocean. In the streets of Chicago, the cacophony of cars and the throbbing intonations of Antarctic icebergs were intertwined for a few days. Perhaps the grating inflections of the White Wanderer's song are still echoing in many heads.

Figure 5 Sound recording of WhiteWanderer Riverside
(Petra Bachmaier and Sean Gallero, luftwerk.net)
soundcloud.com/eartherdotcom/whitewanderer-riverside

The ear before anything else

In frozen worlds, you listen better than you see. On sea ice, you can hear the ice breathing with the sea's swell. The eye, on the other hand, is quickly dulled. The ice invites you to use all five senses at the same time. In this concert, hearing takes over.

Oral traditions make the eye 'subservient to the ear. They define space more by sound than by sight.' But you have to recognise and get used to it. As Edmund Carpenter wrote, we would announce, when faced with an icy expanse to be crossed by sled: 'Let's see what we can hear.' His fellow indigenous friends would say, 'Let's hear what we can see.' The anthropologist goes on to explain:

> The essential feature of sound is not its location, but that it be, that it fills space . . . Auditory space has no favoured focus. It's a sphere without fixed boundaries, space made by the thing itself, not space containing the thing. It is not pictorial space, boxed-in, but dynamic, always in flux, creating its own dimensions moment by moment . . . indifferent to background. The eye focuses, pinpoints, abstracts, locating each object in physical space, against a background; the ear, however, favours sound from any direction.

The ear has a valuable advantage; it does not focus attention. In other words, it does not steal it from the environment by enclosing it in the linear perspectives of a frame. On the contrary, it opens it up to all the clues that present themselves in an undefined context. In the midst of a foggy coastline, Carpenter's friends manage to 'read' the nature around them, the waves that are sculpted by the wind and the swell. And 'loss of sight was not a serious handicap . . . they were not "lost" without [their eyes].'[33] This aspect has already been stressed. We can now deduce that the anthropologist's companions are confident because they know that the audible is less deceptive than the visible. The environment emits sounds and addresses that part of the sensitive body that can listen. This is why official administrative maps, which order the world according to geometric criteria, are sometimes useless in such contexts.

The rhythms of an iceberg's personal life cycle are not always given due attention. As soon as its top is too worn away by wind or temperature, it rotates. Then it rolls over. The submerged bottom becomes a newly revealed top. When it turns over for the first time, the block shows the face which had long remained invisible to us. Its existence is then cadenced by multiple tilts. Let us accept that the cycles of rotation in the water generate a complex sound track.

Noises are made by the birth of icebergs and their subsequent rotations. They are always audible events. You can hear the blocks wheezing as they fall. When they turn over, the wheeze always comes from the impact with the water. But it comes from the depths. So the iceberg presents the base that ensures its buoyancy as a whole and exhibits the part that emits the most sound. Diving oceanographers who reach its point of equilibrium hear the words it utters in the ocean. Faced with this intensive mass, this vital matter, they no doubt consider the iceberg to be an essentially sonorous entity. The realm of the iceberg lies below the waterline. Its soul resides in the underwater world.

If the weather is clear, the spectacle of an iceberg is sometimes surprisingly sharp. This sharpness gives the impression that the iceberg belongs entirely to the realm of immediate appearances. But its invisible and sonorous proportions are much greater. The submerged volumes prevent the viewer who remains on the surface of the water from grasping everything. The eye does not see the centre of what it is looking at. This makes it difficult to understand all the phenomena of iceberg life. Most descriptions of icebergs are surface descriptions. The eye remains on the boat. It does not dive into the cold waters. This is the weakness of the aesthetics of the sublime, which remain, above all, matters of the eye.

Barry Lopez suggests that icebergs completely change our relationship to space. In the sunlight, we can make out the line of the horizon stretching behind their visible forms – some clouds, perhaps, racing across the sky – but there are no landmarks at water level to recalibrate the perspective.[34] It is now clear that this is not the only reason why icebergs introduce confusion into visibility, especially when they emerge from the mist; they scream, they whisper, they breathe. In short, they send out subtle signals that can be deciphered only if you watch them rotate or, better still, if you pass by their submerged flanks.

Part of the mystery of icebergs is undoubtedly the play of light that shines through them. The light reveals unsuspected colours in the inner crevices. But the chromatic variations do not sum up their personality. Otherwise, we neglect the moments when the invisible is embedded in the visible. The philosopher Maurice Merleau-Ponty wrote of painting that 'the very nature of the visible is to have the invisible as its lining.'[35] The same applies to icebergs. On the condition that we admit that the ear awakens our perception of these masses that break up or roll over themselves. It is through hearing that we individualise icebergs. The sounds we hear teach us that visible volumes are connected to depths that contain their biography. It is from below that the voice and the life of the icebergs come up. Each time they turn over or make their way through the liquid immensity, we hear new variations of sound.

Icebergs are not things. When one is confronted with a calving or during a dive, one senses a being living for itself. A whole audible world unfolds, inviting us to blur the boundaries between what is animate and what is inanimate. To think of icebergs as things would be to forget that any body immersed in the ocean becomes a biotic medium. Underwater acoustics reminds us of these simple facts.

As it floats, the iceberg draws us to it. It is strange and familiar. We sense a distant kinship. It envelops us in its long history and invites us to go beyond appearances to discover a world of unexpected lives. It lets us hear it breathing in and out. We have the impression we are swimming next to a whale.

But what are its intentions?

4

Social Snow

It is the end of January 1895. A whaling ship drives through the great ice barrier discovered by James Clark Ross some fifty years earlier. The *Antarctic* and its crew reach Cape Adare, north of Victoria Land, and anchors the ship. They lower a longboat. Seven men board the boat and head for a beach just across the way. When it comes to deciding which one should be the first to set foot on Antarctic soil, geographer Carsten Borchgrevink and Captain Leonard Kristensen have a heated exchange, even grabbing each other by the collar. But the small boat bumps the shore, and a rower, a seventeen-year-old sailor from New Zealand, leaps to avoid capsizing. And there stands Alexander von Tunzelmann on the mainland, his companions watching in amazement.

The original iceberg

This episode regularly illustrates the annals of polar expeditions. It may still be laughed about in specialist circles, when someone asks who was really the first person to set foot on the Antarctic continent. Fortunately, there are many other ways of reaching a shoreline without knowing. At the same time, in Greenland, people used to land silently and do a somersault on the spot, head first. It was customary to do a somersault; it was a way for newcomers to try to convince the local spirit that they had been born there and were returning home. The spirit would be amused enough to accept and protect them.

Today, the Inuit continue to have rites for approaching foreign lands and have a habit of avoiding angering the spirits. Gestures, noises and formulas are all part of arrival rituals. In the Belcher Islands, in Sanikiluaq, for example, one is supposed to 'get down on all fours, like a baby, and go forward while saying, "Ungaa! Ungaa! Ungaa!" to imitate a newborn's wailing.'[1] There is no need to fight; it is better to offer the spectacle of a fictitious birth.

On the pack, the Inuit tell each other stories to pass the time, or just when they feel like talking. They can be tales of drifting icebergs. There are tales of sleds tipping over, of ghastly beings haunting the area, of long waits at seal breathing holes (*aglous*) and of unexpected births. These stories are linked to hunting grounds. They are embedded in myths organised as mental maps. Snow and ice structure social relationships; they are present at most birth scenes.

Paul-Émile Victor and Joëlle Robert-Lamblin relate two ancient practices. In the event of a complicated birth, it is advisable to blow into the stem of a blade of grass and to utter a charm. The aim is to melt the piece of ice (*nida*) that is blocking the exit of the foetus. The term *nida* refers to 'glacier ice, or iceberg ice, which gives fresh water. It is said that the uterus of women who have difficulty giving birth is made of a piece of this ice.' There is another, even more explicit charm: 'You have to get the foetus out, what have you got in there? Is that a piece of ice you have in there? A piece of ice that has turned? You have to get well!'[2]

These charms show that the iceberg is part of the female body. It is her own flesh, her womb. It usually melts and doesn't block the passage of the baby. But, when the labour lasts too long, someone has to help the mother by coaxing the resistant iceberg inside her. He whispers words that will convince the block to dissolve and let the newborn see the world. Related to an essential organ of the female reproductive system, the iceberg accompanies the foetus during its intrauterine growth process until birth. It is then a natural partner in family life, in its games and pleasures.

It is evoked in nursery rhymes for comforting children from the very first moments of life. The following melody is sung: 'He is round, he is shiny, like a little ice block in the water, he jumps, he floats, he plays, like a little ice block in the water, "*Aya aya yêk!*", look up, look at me, little ice block in the water!'[3] The song tells the baby that he is made of the same ice as the piece in his mother's body and that this piece is drifting in the ocean, far beyond the village. Both behave in the same way. As unpredictable as an iceberg, the child bends over, nods and turns over, making a noise. His eyes sparkle with a thousand lights.

Years later, the child begins his shamanic initiation. The young Inuit is now a teenager. He must become clairvoyant and access the invisible world that surrounds humans. He lets himself drift for a while on a piece of iceberg or an ice floe. If he manages to return to his people, it is because he has been helped by spirits. The proof will be that he has negotiated with them and that the spirits have passed on some of their knowledge. The iceberg is always there with him, helping him through the stages of his life.

Ice does not only surround humans. It vibrates within them, it lodges itself in their deepest intimate places, to the point of reversing perspectives in a radical way. Anthropologist Bernard Saladin d'Anglure evokes the testimony, more recent than the charms quoted, of an Inuit man in his sixties:

It will be hard to believe what I am about to write: I can remember before I was born. It seems like a dream. I remember I had to go through a very narrow channel. The passage was so narrow I thought it would be impossible. I didn't realize the passage was my mother – I thought it was a crevasse in the ice. That ice crevasse must have been my mother's bones. I remember it took a long time to go through. Once I turned back – it was too hard. But finally, I was outside; I was born. I think I opened my eyes inside my mother but after I was born I opened my eyes again and all I could see were two little cliffs on either side of me. I often remember this: I saw something blue and those cliffs which were exactly the same . . . which were probably my mother's thighs.[4]

This intrauterine memory is striking. It combines two patterns. An adult remembers the moment of his own birth. First point of view reversal: he goes back in time and becomes the foetus he was. The description that follows goes even further. It tells us that he leaves his mother's womb like an iceberg leaves its glacier. Second inversion of point of view: the witness describes his coming into the world in terms of a glacial calving.

The being making his way towards the exit walks through a corridor of cracks. He is conscious. He grasps everything that happens to him. He knows that his mother's body is that of a glacier. Inside the white mass, he compares the ice to a skeleton. The crevices represent the parts of a skeleton that are working and creaking, the joints of a whole body in full labour. It's a slow process. The foetus moves backwards, like a block that detaches itself from its 'parental glacier' (as the glaciologists say) without really breaking away. The calving stops for a while. Then the movement resumes.

Once released, the newborn floats as if it had fallen into a bay between small ice cliffs. These cliffs correspond to the outer body of its mother. The blue colour even provides a clue to age. In an old glacier, it indicates that the air bubbles in the crystals have disappeared under the pressure of their accumulation. Only part of the light is reflected, the blue part of the spectrum. If a man remembers this colour, he can deduce that his mother was not very young when he was born.

The testimony of this Inuit adult mentioned by Saladin d'Anglure reformulates the 'perspectivism' of Amerindian cosmology. In the work of anthropologist Eduardo Viveiros de Castro, this cosmology presupposes an original state in which humans and animals communicate by sharing the condition of humanity and exchanging their points of view.[5] There is apparently nothing of the sort going on here. The original condition is that of ice and the viewpoint of the foetus is that of conscious matter. If the American, Danish or Greenlandic fishermen of the nineteenth century predicted the birth of an iceberg as the appearance of a whale calf, the testimony quoted here shows that the birth of a human child is sometimes understood in terms of an iceberg calving. So, the mother herself is a glacier.

To recap: how do humans perceive themselves in this world order?

The witness's recollections are indeed those of a man. But they bring back the moment when he was still a piece of ice. Then a metamorphosis takes place. As it is born, the baby iceberg is transformed; it leaves its icy home and takes on the appearance of a human being. Initially frozen matter caught up in the event of its birth, the human gradually tears itself away from its non-human state.

A few general conclusions may be drawn from this brief account. Firstly, that the division between animate and inanimate beings is not operational here. On the other hand, that the point of view of the birth is focused on the relationship between the iceberg and its glacier.

Perhaps this is why the human infant retains a lifelong kinship with the iceberg, even to the point of wanting to imitate it later in its games with other young adults.

Saladin d'Anglure observes that, in Canadian Nunavut, individuals from different places express similar intrauterine memories. In each case, the foetus is conscious. It lives in symbiosis with its mother, deep in the intimacy of its ice cave. It shares all its emotions, from sadness to joy. It hears what is being said in the outside world and anticipates the dangers. This type of reminiscence seemed to be quite common at the time the anthropologist was conducting his research. It is part of a 'narrative genre' that the shamans reactivate when they have to decide the fate of dead souls.

Inuit cosmology is very consistent. The souls of deceased people travel through the orifices of the body. The uterus is located in this distributive world, the organ playing the role of a space of transition. Any soul that plans to be reborn must find a way to return to the world of the living. The preferred route is through the crotch, which is 'a way in for the soul and a way out for the fetus'.[6] The Inuit have a theory that applies to every case. Every person has

> a double-soul (*tarniq*), a miniature image of oneself, which is encapsulated in an air bubble (*pudlaq*) and lodged somewhere in the groin. Everyone also has a name-soul (*atiq*), a psychic principle inherited from a dead person or a spirit, which encompasses all of the experiences and abilities accumulated by everyone who previously held that name. At death, the double-soul escapes from the bubble and expands to the dead person's size, becoming his or her ethereal replica. It will then roam around the grave until the name-soul manages to reincarnate. The two souls then part company: the double-soul goes to live eternally in the hereafter, looking like the person did just before death, whereas

the name-soul lives again in a fetus and enters a new cycle of human life.[7]

Each birth is an opportunity for the name-soul to take shape again. An entire lineage is established in the mind of the baby that emerges into the light. The closing scene of the film *Atanarjuat: The Fast Runner* (made in 2001 by Inuit filmmaker Zacharias Kunuk) makes this point clear. A child appears at the entrance to the family igloo, as if just born, as the name of the recently deceased shaman is called out. The small community decides that he will bear the same name. It revives in him the name-soul of the deceased. The dead person may be a member of the family or related to a group of friends. The newborn child may have different names. Several name-souls may take on a new existence in his body. Each child is thus the child of a group before being the child of a particular mother and father, the heir to the peregrinations accomplished by ancient souls. No birth is a clean slate. Every life that begins is a life that continues.[8]

In a chapter of her novel *De pierre et d'os* [Of stone and bone], the writer Bérangère Cournut brings up this theory of souls and usefully includes the role of icebergs. The heroine, Uqsuralik, follows, in a kayak, Tulukaraq, the son of one of the families who took her in after she wandered onto an ice floe. They both arrive at the foot of several drifting masses. Tulukaraq knows the icebergs well. He can tell the young ones from the old ones by their smooth or striated sides, which are sometimes even lacerated. He knows that they become unstable when the sun has heated their tops or when currents, warmer in summer, have eroded the submerged parts. They walk along them in silence. Then they hear the sound of flowing water. Paddling their kayaks, they slide into a recess. They want to work out where the run-off is coming from.

The iceberg is a labyrinth. It is a world of communicating cavities and channels. From the outside, it appears to be a compact, homogeneous block. On the inside, the two friends discover tortuous passageways. They begin to think they could get lost in it, as in the passages of an enchanted castle.

Suddenly, the colours are shimmering so much our heroine is stunned. A light blue dominates; time is suspended. Sounds dissolve into their echoes. Light itself floats. The walls are transparent. Tulukaraq's kayak is fading behind a column of ice. It appears, then disappears. Uqsuralik stops her boat. She can no longer see him. But in the muffled atmosphere she hears his voice echoing in the cave, filling the maze.

Her friend is asking her to join him near the stream. First, he invites her to look up so that she can see the holes in the top of the iceberg. Thus begins their conversation, at a distance, about the fate of souls:

So now I'd like to ask you a question, Uqsuralik: do you know why the dead sometimes end up in heaven, sometimes at the bottom of the sea? . . .

I don't know, Tulukaraq.

Well, because they go through holes like the one above my head, Uqsuralik.

What do you mean, Tulukaraq?

Well you see, Uqsuralik, if you were to die right now, your soul-name – your *aleq* – would escape through a hole like this. So you would join the dead up in the sky, waiting up there to come back to live in a body on earth.

What if I died at another time, Tulukaraq?

Well, if you died after a storm, Uqsuralik, or if you died after the summer, you would go through the same hole, but you would end up at the bottom of the sea.

And why is that, Tulukaraq?

Well, because in the meantime, Uqsuralik, the iceberg would have split and turned over! Going through the

hole that is now at the top of this wall would lead you
on a journey to the bottom.
 But then, Tulukaraq, what would happen to my *aleq*?
 Hmmm . . . it would probably seek refuge in Sedna's
hair to become wild ocean animals, Uqsuralik.[9]

Tulukaraq is not a shaman. But their journey is like an
initiation. The heroine is astonished that he spoke his
'birth name' for the first time in the iceberg cave. It is
a ritual word. Her companion is preparing her for the
question he then asks about the dead. He is anticipating
their discussion on the significance of the opening of
the iceberg. Such breaches do exist in reality, although
more often they affect the sides. They are created by air
bubbles escaping underwater and scraping the entire
mass as they rise to the surface. When the iceberg tilts or
crumbles, the areas weakened by this underwater work
leave gaps of varying size.
 The opening at the top of the iceberg is a sign, for
our wanderers, of a passage of souls. It is another figure
for the maternal womb. The two protagonists imagine
their death and the metamorphosis of their soul-names.
Uqsuralik's would fly away and pass through. It would
return to the sky to wait among other souls. It would
wait for an available newborn body to restart its cycle
of existence and perpetuate itself. If the iceberg turns
over, the hole would be facing downwards. Uqsuralik's
soul would then escape into the ocean. Like all sea animals, it would live in the hair of Sedna, the goddess of
the waters, and would end up looking like one of them,
perhaps a seal or a narwhal.
 Bérangère Cournut has used the ethnographic archives
to inform her novel. These support the mythological
representation of the iceberg as a tunnel of rebirth. Souls
circulate through it, change state and live again. The
scene in question depicts a disruptive initiatory ordeal.
A blinding whiteness; the two companions paddling in

the cave losing sight of each other; the ice getting in the way. Tulukaraq's voice becomes that of a benevolent spirit who warns that the iceberg may turn over.

The young heroine understands the message sent by her companion: the fate of the soul-names depends on the iceberg's own life. The range of human metamorphoses is articulated with the iceberg's life cycles. The rotations change the set of possibilities somewhat, depending on whether the exit points to the heights of the sky or the depths of the ocean. They make reincarnations more hazardous: sometimes human, sometimes non-human.

Tulukaraq is right to warn his friend: 'an iceberg is a world that can topple over at any moment.'[10] As he says this, he reappears, as if by magic. The initiatory episode is complete. Before they leave the cave inside the iceberg, Uqsuralik asks Tulukaraq what would happen to him if the iceberg turned over. He replies that he would probably reconstitute the 'first islands' if he had to escape by sea.

This last detail confirms the apprenticeship value of the episode. An immemorial time is evoked – that of the islands. Icebergs were the places of the first human settlements. Bernard Saladin d'Anglure recalls, following Franz Boas and Knud Rasmussen, that, in the Inuit representation of the world, 'the habitat of early humans, a place where islands floated on the sea and could tip over.' He adds that 'The 1930s saw the beginning of active Christianization among the Iglulik Inuit, but they still saw the earth as a flat disc surrounded by the sea, with everything tenuously balanced on four pillars standing over a lower world. Upon the earth stood four other pillars that supported the firmament and the upper world.'[11]

In this cosmology, icebergs are the original islands, the blocks in an archipelago of initial sociability. Then there comes a time when they are overloaded with humans; they tip over and sink. Humans fight wars, they lose their immortality and the great migrations begin.[12] The anthropologist's analysis is based largely on oral testimonies. Almost word for word, it echoes, in a surprising way, the fantastic account of the Christian journey of Brendan de Clonfert, which we recounted in the first chapter of this book. In both cases, the world rests on pillars. And these pillars are icebergs. They wobble a lot. They have to be put back in line all the time. The universe is frail. It is unsteady on its legs. The shamans say that it is liable to pivot on itself, as happens with all icebergs.

We can see that icebergs play a major mythological role. They are associated with invisible kingdoms populated by spirits. As ancient island territories, they are the first refuges. They determine human births and help dead souls to be reincarnated. They are the original characters, the foundations of the world.

Practical words

In Antarctica, the smells of the snow are multiple, its textures infinite, its crunching polyphonic. But the seasonal variations are small and extremely restrictive. The winds blow the snow away and the years pass on the frozen continent at the rhythm of alternating light and dark, depending on the area. It is perhaps less easy to say, as Mario Rigoni Stern does, 'I have plenty of snows in my memory.'

The writer was born in 1921 on the Asiago plateau in Veneto. During his childhood, Cimbre (a German dialect) still recorded all the types of snow in the different months of the year: it was weary in autumn (*brüskalan*), abundant in the first weeks of winter (*sneea*), wet or resistant as spring approached (*haapar/ haarnust*), restless in March (*swalbalasneea*). Then it became light in April (*kuksneea*) and scarce in May (*bàchtalasneea*).[13]

In the far North, sea ice and icebergs enrich language. Words retain their sounds, their pace, their movements, sometimes even their moods. In Qaanaaq, Greenland, the term *ihittop* refers to the iceberg that falls and disintegrates in the water; *oq"rad"dlattoq* to the beginning of its rotation; *itsineq* to the submerged part; *putad"dlartoq* to the ice that bursts through the surface of the sea. In Kangiqtugaapik, on Baffin Island, the word *sirmik kataktuq* denotes the iceberg that has just been born; *niiqquluktuq* the ice that has been broken by the waves and whose agglomerated pieces produce a grating sound; *qinuaq* the anxious ice that wanders for a long time; *aulaniq* the ice that comes from the north and is always slipping on the waves. Finally, in Barrow, on the north coast of Alaska, the term *qimmiaqrugauraq* is used to describe the noise caused by the accretion of bits of sea ice. The sound is reminiscent of a puppy calling for its mother.

The cases could be multiplied. Anthropologist and linguist Louis-Jacques Dorais points out that Inuktitut is an 'agglutinative language'. Its words are made up from a root that gives the initial meaning. Then other elements are added that twist the meaning. Neologisms are created. For example, from the word *siku*, a generic name for ice, the word *sikuaq* refers to the first layer of thin ice that forms in the autumn on pools of water, while *sikuliaq* refers to new ice on the sea or on stone surfaces.

This proves that everyday language nourishes the experience of reality. The more we use it, the more we become aware of the proliferation of phenomena. There is no general theory here, only distinct circumstances that are reflected in the statements. And the more precise the language, the better it defines what is said about the states of snow or ice. What is said is this: everything changes, nothing is static. This is why words tell speakers how to act. Depending on whether the ice breaks after he has tested its strength with a harpoon (*qautsaulittuq*) or whether it cracks due to changes in the tide (*iniruvik*), the hunter will choose to go out on the sea ice or to stay at home.[14]

Geographer Shari Gearheard explains how the vocabulary of the Inuktitut language helps hunters to orient themselves in their environment. The term *qisuqqaqtuq* refers to the type of snow surface that normally covers part of North Greenland in June, during one of the phases of the moon. It melts during the day, forming a film of water. The refreezing of this water during the night makes a thin, crunchy 'top surface'. Armed with this knowledge, the hunting Inuit wait for the moment when they can take out their sleds. But climate change is changing the traditional patterns. This snowy state now occurs in May.

Also, with the month of March comes the time when everyone normally follows the tracks they have been using every year for decades. This is now a risky business. The currents are too strong and the ice cover too thin. The sled dogs are sinking their paws in. Cavities in the pack ice can cause the sleds to tip over.[15]

Words express the strength of the environment and its particularities, just as they testify to the human ability to take advantage of it. Edmund Carpenter explains that language is the tool with which the Inuit transform their practical and immediate environment into a human world:

> they use many 'words' for snow which permit fine distinctions, not simply because they are much concerned with snow, but because snow takes its form from the actions in which it participates: sledding, falling, igloo-building. Different kinds of snow are brought into existence by the Eskimos as they experience their environment and

speak; words do not label things already there. Words are like the knife of the carver: they free the idea, the thing, from general formlessness of the outside.[16]

Inuit make little distinction between nouns and verbs. All words, says Carpenter, are more or less forms of the verb 'to be'. They do not, however, designate realities that already exist. They bring them into existence. Speaking is a way of animating the world. It is a new birth every time. It has already been suggested that a child's name is given to it at the very moment it appears and its mother takes it in her arms. Its birth interrupts the list of eligible names often being mumbled by an old woman. In collective representations, naming and coming into the world are two events that the Inuit strive to make contemporary.

Language embraces the environment. It has followed the seasons for centuries. The cyclical nature of language means that words evolve with the 'conversation' that residents have with their lived environment. Here, as elsewhere, they 'give their attention to varieties of space.'[17] In every discussion, landscape is an inevitable third party. It cannot be excluded. It is already in the language and precedes the speech. So that every address to a person is an address to the territory and the natural elements that constitute it.

All these examples confirm that language constantly registers the modified states of ice. The law of its operation is one of fundamental biocentrism: humans live with ice; they literally incorporate it. As it is part of their daily life, it inhabits the words of practical language. Ice contributes to the development of collective norms. It is at the origin of a mental dictionary that contains all the rules of an art of caring.

The sun was shining outside. It was warm for this month of the year. The afternoon was stretching out

like a great lazy beast. Silence reigned in the small room
of the Yvonne-Oddon research library of the Musée de
l'Homme in Paris. I was consulting books on the glacial
vocabulary of the Inuit when, suddenly, a ballad by
Gustav Schwab, *The Horseman and Lake Constance*,
came to mind.

A man is riding in haste. He is riding through
Switzerland and has only one thing on his mind, to
reach Lake Constance. He arrives at the top of a hill.
A plain blanketed in snow stretches out before his eyes.
The winter is freezing, fog is forming. He sets off. He
hears snow geese calling; a moorhen tries to fly away.
No villager is lingering on the side of the road to show
him the way. He continues until he can see trees emerging from the fog, houses and hills in the distance. When
he arrives in the town, he asks a young girl leaning on
a windowsill where the lake is. She tells him that the
lake is right behind him. She is amazed by the luck of
this man who emerged unscathed from the night. He
has made his horse's hooves sound on the ice yet has
not disappeared into the depths of the lake. The people
of the village gather around the rider. They offer him
bread and fish. He is dumbfounded, understanding that
in crossing the lake he has run the worst of dangers. His
vivid imagination makes him hear the thunder of cracking ice. He sees himself on it. Seized by fear, he falls off
his horse and tumbles to the ground, dead.

Gustav Schwab's parable applies to all icy countries.
One does not move onto a frozen surface without first
observing it in great detail. There is another truth here;
on the ice, one can never be too careful. The Inuit say
that the sea ice arouses mixed emotions. '*Ilira* is the
fear that accompanies awe; *kappia* is fear in the face of
unpredictable violence.'[18] Being confronted with a polar
bear belongs to the first type. The prospect of sliding
on a sled over thin ice is of the second type. In the end,
the sea ice hides more perils than an encounter with a

bear. Even the vigilant hunter, who engages in 'accurate climatic analysis', may be surprised by changes in the snow surface that he has not anticipated.[19] On the pack, nothing can be taken for granted. Everything has to be negotiated. This is why ice is a collective affair.

The sea ice as an institution

Icy worlds are full of mischievous beings that fly through the air. Each is king in his own domain. The sharp ice ridges that adorn the snowdrifts formed by the wind have their own spirit. The more the blizzard blows, the more it blusters. It brandishes its fists, bursting into laughter, and treats the hunters to powerful snow flurries. Humans are forced to be more inventive in devising rituals of approach.[20]

When the ice floes press against each other and create cumulative thickenings (or hummocks), it forces the villagers to stay in their homes and prohibits them from moving. Then they go to the cemetery and place in front of one of the tombs 'a man's skull with seaweed for hair and a block of ice about the shape of a skull with creeping plants for hair.' Then they intone the following spell to chase away the ice, repeating each phrase and concluding with a long breath: 'One leg is bent (*twice*), the other leg is stretched (*twice*), the intestines vomit (*twice*).'[21] The charm dramatises the struggle of the land with the sea. It says that men want to prevail over the ice. Their breath is meant to keep the hummocks away and free the waters so they can use their kayaks. These rituals and customs prove to the ice that humans want to negotiate with it and that they need to perform certain tasks to survive, including hunting.

Such rituals are not enough. Living on sea ice means never being able to stop looking at it. When he goes out, the hunter first identifies where the snow is deepest by noticing areas that are sometimes bright, sometimes dark. He knows that the ice is thinner where it is isolated by snow. When he navigates through chunks of pack ice, he sees reflections over the sea in the fog. The black lines on the horizon are 'water skies' that indicate open areas of ice.

The ice cover may be thin even though it appears high due to the tide. If temperatures remain constant, neither the density nor the compactness of the ice will change. If temperatures rise, however, the thinness of the frozen surfaces will become dangerous. In Greenland, 'bearded seal-skin ropes' (*ugjugajataktunaat*) are used to indicate the texture of the sea ice. The nature of the weave of the ropes changes just before the temperatures change. When cracks form, hunters know that the pack ice will crack. It does not matter how much the ice height changes with the tides. The seal skin sends people danger warnings.[22]

Birth memories, mentioned above, already show that there is no real border between mental, social and physical facts. The further we go, the more we realise that ice is integrated into the fabric of the institutional world. It is part of a set of interdependent relationships, norms and meanings shared by groups. It controls many of the actions, thoughts and customs among people. In short, it helps to organise human 'life forms'. It alone encapsulates the 'psychological background of needs, desires [and] natural reactions and [the] historical background of institutions and customs' that enable individuals to evaluate their intentions and decisions.[23] When someone plans to go on a trip on the sea ice, they summon a system of laws and customs in which the multiple states of the ice are recorded. They know what to expect.

Joelie Sanguya is from Kangiqtugaapik (Clyde River), Nunavut, Canada. He writes that sea ice is home in both a permanent and temporary sense – we think about it, we talk about it, dream about it and, as Jacopie Panipak said, we are *homesick* for it when we are away from it for too long. It also provides a temporary physical home sometimes, a place to set up our hunting, whaling and fishing camps over the seasons. In the past, before settlements, we would live on it during certain times of the year. It is the sea ice that in the past was the ultimate source of all warmth – a word that Elders today use frequently when they reminisce about their lives growing up and living on the ice. They remember life in the igloo, set on the sea ice for its superior warmth over the land, and the warm glow of the *qulliq/qulleq* – the oil of this lamp from the animals hunted from the sea ice. Sea ice *means* warmth, and the light of home. Today, the sea ice still evokes profound feelings and playful moments. Not only is it where our food comes from, it provides a place for Christmas games, New Year's fireworks, soccer games, local dog team races and family picnics. It is where life happens.[24]

This testimony is crucial. It confirms that the sea ice is experienced as a social institution and that it develops a collective mode of existence. Some remarks may seem counter-intuitive to those who do not live in this environment. How can you associate the ice with the warmth of the home? But a well-built igloo, with the right kind of snow, maintains a satisfactory ambient temperature. And the oil lamp has the role of diffusing a low level of light that relaxes the body as well as the mind after a day of intense activity. The rhythms of outdoor and indoor living define this conviviality; one constantly moves from the cold to the warm.

The sea ice opens the travel season. And travel brings friends and family together. When they stop to camp, the long nights pass better; we share stories about our ancestors, we rediscover games, we sing old nursery rhymes, we get up to dance for fun, we also ask about the younger ones we haven't seen for a few months: have they grown up well, are they good students, what will their future jobs be? The sea ice means that people spend time together. It acts as a 'bridge'. As a symbol of association, it brings together people who are often separated by the limitations of the natural environment. Everyone takes the opportunity to rediscover the old tracks and check that they are still passable. As for the new ones, they quickly provide opportunities for stories that are passed on from generation to generation: 'Travelling sea ice is a part of our individual and family stories and histories, which we continue to tell and add to,' says Joelie Sanguya. The sea ice creates society. It brings people together and maintains ties. It holds people together in a harsh environment. In two words: it restores and strengthens sociability. The ice pack is the true homeland, the other name for freedom.[25]

On the way, however, care must be taken. The frozen expanses like to surprise. In concluding his testimony, the same resident recalls with emotion that one of the village's grandfathers once wandered out onto the ice. He had set seal nets and wanted to make sure they were in good condition. But he never returned. He was probably 'carried away as the ice broke off and drifted away from land.'[26]

Even a seagull hunt can be very dangerous. At first glance, one thinks it is ingenious to hide in an igloo after putting bait on the roof. Pinaitsoq, one of the characters in the authentic Quitdlarssuaq epic, did not anticipate the break-up of the sea ice. Even before he could collect his prey, he was exiled, drifting on the waters.[27]

Local chronicles are full of sudden disappearances. Hunters leave, never to be seen again. They appear only in the tormented dreams of members of the community.

Sensitive colossi

The piece of ice carrying the grandfather in Joelie Sanguya's story must have broken free because of an unpredictable cleavage. However, in other stories or testimonies, blocks may move away for different reasons. If humans do not behave decently towards it, the ice does not hesitate to punish this behaviour.

A short story is enough to demonstrate this.

Once upon a time, a group of hunters decided to go and quench their thirst on an iceberg. They climbed to the top to collect its meltwater. It was fresh and abundant. The ocean was shining brightly. Nothing on the horizon. They were alone. Then one of them started to urinate. When they noticed, his companions were frightened. They tried to stop him, but they were getting in each other's way and slipping on the solid surfaces of the iceberg. They were desperate, afraid of waking a monster, an angry behemoth. What followed proved them right. The iceberg began to move. It didn't wait for the poor bastard to finish. It cracked all through and quickly toppled into the dark water with a loud roar. The hunting party was swept away by the waves.[28]

The punishment is severe. Everyone drowns because of the fault of one. It must be that the mistake was inexcusable in the eyes of the iceberg. The blocks of ice are in any case irascible pranksters, mischievous beings who play on the nerves and with the very lives of humans. The indigenous peoples know that they are both precious and feared acolytes, experts in metamorphoses, sometimes very sensitive. They find them up to their tricks in legends that frighten adults as much as young children.

The explorer and ethnologist Paul-Émile Victor reported a provocative anecdote from the time he spent with his Greenlandic hosts:

One evening in Oumivik, Kara, then a young girl, and her friend Odîné went out in the night to fetch water from the stream. On their way back, their attention was suddenly drawn to a huge iceberg, moving alone among the other ice in the fjord, which was motionless: a white mass, almost phosphorescent, almost alive. While Odîné, immobilised by fear, could not take a step, Kara rushed into the hut to tell what they had seen. Immediately, some men came out with their rifles. But the strange ice cube had already disappeared 'under the earth, under the house'. – It was one of those ice *toupidet*, one of those *'poubik'* that often crush houses.[29]

Toupidet are monstrous, and usually invisible. They live everywhere, in the sky, at the bottom of the ocean or on the ground. Shamans, the *angakut*, sometimes talk to them, suggesting that they do one thing or another. They meet them during their travels in the world beyond after they have flown away, bound hand and foot, behind the curtain of the tent where they hold office. The *toupidet* are created by the *idizitsut*. The latter are a special kind of shaman: they bypass traditional ceremonies and communicate with the spirits in secret. They are clandestine. They collect materials, mix them in a thousand ways and blow on them to give them a silhouette. The beings they make are hybrids of non-humans (the head of one animal and the body of another). Sometimes the bodies unite parts of a non-human with those of a human. These shamans are 'getting back' at something. They seek to 'harm'. Their *toupidet* have abnormal attitudes. They go out at night and prowl around like ghosts.

In Paul-Émile Victor's anecdote, the *toupidek* (the *-dek* suffix marking the singular) is an iceberg. The gang of friends luckily escapes the threat. Perhaps the guns frightened the spirit that had taken on the appearance of a block of ice. Perhaps the desire for revenge had no real or sufficient basis. When the *toupidek* misses its target, it has what is known as a 'return of the curse'. It turns against the shaman who shaped it. In this case, the iceberg that disappeared will surely have decided to go and crush its demiurge's house to dust during its sleep.[30]

This anecdote shows that icebergs do not only set the norms and values of a community through narrating events such as births or in practical everyday language. They also transmit strange stories that enrich collective legends. The ice imposes its law. This is a feature of human societies in other latitudes and natural environments.

The writer Charles-Ferdinand Ramuz drew inspiration from folk tales to write his novel *La Grande Peur dans la montagne*. In this book, published in 1926, the characters are in fear of glaciers, which move like nimble and lively beings. A mass that moves forward, while giving the impression of being motionless, is bound to feed rumours. In the valley, two groups are at odds. The reason for the dispute is a meadow that has been abandoned for twenty years. The village elders are convinced that their alpine pasture is 'bad' and that ice in the mountains is covering fields that were once cultivated. The younger people want to take advantage of this. But a disease quickly infects the herd they keep on the slopes. Quarantined, the workers are cut off from the world below. They neglect the cows. One of the characters, Barthélemy, ends up milking them. Milk spills onto the ground, yet another desecration.

Tired by so many trials, Joseph, the protagonist, wants to go back down to the village and see his fiancée. He is forced to climb up to the glacier to bypass a dam. Confronted by the mass facing him at this moment, he does not dare turn around, showing it his back. He is afraid that this furious animal that destroys houses and herds will 'really start moving and leap on him from behind.' He is paralysed. The glacier stands in his way. When, at the end of the novel, Joseph enters the lunar world of the high snows, vapours surround him. The shapes of the landscape become blurred and doubled. Ghosts appear. He sees the corpse of his fiancée and comes across a lone hunter who never stops laughing. Panicked, he shoots at the glacier. Then the glacier coughs as if it had asthma. It cracks all over and releases an avalanche.[31] In the novel, the glacier reacts in this way because of bad things being done to it. It is as if the sacred heights must remain inaccessible to humans, if they don't want to be turned into eternal wanderers.

Ramuz's text could be reduced to a series of superstitions. We could fall back into the age-old quarrel between the faculties of reason and imagination. The former establishes the correct causal relationships between phenomena. The second erroneously imputes intentions to material entities. It is a sterile opposition. What is essential lies elsewhere, as you may have guessed. It is to identify the effects of social structure. In many parts of the world, it is believed that glaciers get angry when offences or aggravations done to them are not acknowledged. Better still, their anger reveals a dysfunction within human communities.

Back to the Great North, this time among the Tlingit and Athapaskan peoples.

In Alaska, one clan recounts that a lagoon appeared behind a glacier because someone killed a dog and threw its remains into one of its crevices. Humiliated, the glacier left Yakutat Bay. In a different version, another clan knows that brothers went seal hunting in the same bay. The boat was lifted up by the waves caused by an iceberg breaking up. They ended up in the mountains, in the cold. One of them died.

The glaciers are home to countless clan stories. There is another theme in the stories from this region, that of the 'woman in the glacier'. It applies to situations where settlements were threatened by the rapid advance of masses sliding, because of their own weight, along mountainous slopes. A woman stays to help save all the children. She thus allows the clan to escape quickly. This character is important. She is used to measure the effectiveness of collective escape tactics in the face of a threat. The young woman would be ashamed if she had to survive the disaster when others died. For her part, an elder of the Tlingit Indian tribe, Elizabeth Nyman, confides that the retreat of the Taku glacier can be explained as follows: a slave died there and drops of blood have beaded on snow.[32]

In her stories, anthropologist Julie Cruikshank notes that glaciers are irascible beings, sensitive to all humiliations. One of her interviewees, Annie Ned, says they can't stand the sight of blood or bad smells. They have great olfactory capacities and are loath to smell the funk of old clothes or dirty blankets. They are, however, attracted to cooking odours, just like bears.

This last motif is a recurrent one. The anthropologist's storytelling friends insist that 'cooking with grease' near a glacier is strictly prohibited. They are always wary of fried bacon. They list the rules to be observed: '[food] should be boiled and never fried, and none of the grease should spill from the soup.' Cruikshank takes note of these prescriptions. She concludes that only 'circumspection defines appropriate behaviour'. When she tries better to understand the reasons for the ban, she is always told that 'cooking with grease causes the glacier to come out'.[33]

There comes a day when the anthropologist takes a closer look at the stoves used. She finds that bear, elk or sheep fat solidifies. It turns white as it cools and starts to look like a block of lard. Such a phenomenon evokes a mass that is both rigid and viscous that cracks when heated – in other words, a glacier that breaks or even explodes. She understands that the kinship of these features is enough to make certain culinary practices taboo.

What is the meaning of these narrative arguments? In her work, Cruikshank shows that glaciers have a specific social function: they 'redistribute areas of responsibility' among clans. The latter 'pay' to survive in the area. Glaciers sometimes react to humiliations by killing some of the people. The effects of their anger also push the clans to change location, to disperse over the territory and to distribute themselves differently.

Glaciers live, they move. They intervene in the social life of indigenous peoples and even 'listen' to their conversations. As if they were looking for evidence of 'human *hubris*' in order to raise their anger and eliminate people. The *hubris* in question is neither the sign of overreached human pride nor the sign of implacable fate. These clan stories are not Greek tragedies. The *hubris* is more of an affront. In the accounts evoked by Cruikshank, the affront does not concern only the victim but rightly reverberates on everyone: the shame felt by the glacier is felt by humans. The disrespectful acts of a few individuals, who do not obey the bans, hurt the dignity of the glacier. They also tarnish the honour of the community. *Hubris* arouses collective shame. Anger is shared.

Any violation of social imperatives calls for punishment. The glacier takes care of it. Its fury presents the thoughtful image of a threat that is inside the community of humans, as a warning. The glacier suggests to the villagers that they are the victims of insults for which they are not all responsible. The insolence of the few creates a guilty conscience among the others, who feel responsible for what the glacier has suffered.

The catastrophes that glaciers cause are judgements. These judgements are understood by populations who strive to change the attitudes of their members by asking themselves, for example, how they can correct the mistakes made. The role of glaciers becomes explicit: they push the community to tighten up on the standards that organise it. They remind it of its elementary social rules by showing the dramatic consequences of their transgression.

Cruikshank emphasises that the glacier thus maintains 'mutually constitutive relations' with humans in terms of both practical and moral life. It punishes, of course, and some 'negotiations' with it 'fail'. It nevertheless helps residents reflect on the conditions that will allow them to alleviate their feelings of guilt and to amend their behaviour. The hierarchy between the non-human entity and the humans is functional here. It allows individuals to organise their perception of the world by integrating unexpected natural events. What, in the end, is a sensitive glacier? It's a glacier that gets angry because the rules specific to a universe of which it is entirely a part are not followed.

'We are sleepwalking'

For those who negotiate with natural entities as if with partners, the kinds of behaviour considered irresponsible for humans are of several orders. Not only urinating on an iceberg or frying meat near a glacier, but also bulldozing a hill to lay a railway line, diverting a river because of a gas pipeline, submerging an entire valley in order to build a hydraulic dam. When disasters strike, many indigenous people say that nature gets angry. It causes devastating storms, floods or avalanches. Nature manifests itself loudly because it is invested with moral power.

Anthropologist Ann Fienup-Riordan and fellow oceanographer Eddy Carmack have worked with the Yup'iks of western Alaska. From them we learn that, for coastal residents, the sea is a part of *'ella'*. This term evokes the weather as well as the world, the universe or consciousness, depending on the contexts of speaking. Nowadays it is used to talk about the atmosphere, the environment or the climate. In short, it describes an ecosystem in which the two notions of 'nature' and 'culture' are so closely interwoven that they have no separate meaning. Fienup-Riordan then quotes Paul Tunuchuk, a resident of Chefornak, who exclaims, a little annoyed by what he sees in his community: 'instructions attached: air, land, water. And they [the ancestors] mention that we must treat it with care and respect. What will become of us if we do not treat it with care?'[34]

Many older Yup'iks think of the ocean as a conscious and sensitive being. They are convinced that the sea is never unaware that a member of the tribe has acted at some point in a reprehensible manner. Since 'the ocean has always been aware and knowing', everyone must guess that they are no longer allowed to go to sea for a while. Another example: the word *eyagyarat* refers to traditional abstinence practices after birth, death, illness or miscarriage. If a child dies and one of the family members goes hunting or fishing before the required period of abstinence has expired, which applies to everyone, tumultuous waves will shatter his boat into a thousand pieces. If he doesn't go, he'll have to wait anyway until the marine mammals move on their own. This is how the ocean behaves: 'That is the *piciryaraq* [way] of the ocean . . . *nallutaituq* [he knows everything that is going on].'[35]

This conception insists as much on the relationship of humans with non-humans as on the relationships among humans. To put it another way, the way humans view each other determines their relationship with the beings that inhabit wilder worlds. Not understanding this upsets all interactions. In the same interview, Paul

Tunuchuk argues that contemporary climate upheavals are linked 'not just to human action – overfishing, burning fossil fuels – but to human *interaction*.' He explains that members of his community find themselves in this situation today because they are 'not being dutiful to each other. It's as though we are sleepwalking.'[36]

His position is clear and others share it. Finding solutions to global warming requires correcting oneself and waking up. It's not just about changing one's actions. We must open our eyes and reconnect with certain old rules of life. Reducing carbon emissions is necessary. But that will not be enough. It is necessary to reform souls so that they again take the path of collective values. For many, this is the only way to stop meteorology from lying. The people Ann Fienup-Riordan and Eddy Carmack have met are struck by the volatility and unpredictability of winds. They lament that *ella* herself has become a 'liar'.[37] They believe that changes in ecosystems follow the evolution of reciprocal behaviour within the human species. The *ella* entity is an 'intensely social' figure. It reacts to the actions of humans towards other humans and thus verifies one of the community's adages: 'the world is changing, following its people.'

Agreements with non-human beings were made a long time ago. These agreements are the result of negotiations carried out by the ancestors. They have been internalised by subsequent generations over the years. Today, the inconstancy of the climate partly reflects the non-observance of traditional rules by members of the community. This is what the person speaking to Fienup-Riordan and Carmack means. Invoking the rules is not necessarily an anti-modern attitude. It's more about being consistent. If the whole of society has been framed for ages, it makes sense that the melting sea ice and the weakening of customary bonds occur together. Paul Tunuchuk does not enquire into any hierarchy among causes. He fully recognises the anthropogenic impact on the climate and adds that human relationships matter too.

The attribution of a moral failing is explained through a system of thought and action where glacial entities are supposed to assess the degree of organisation or social disorganisation of human groups. If we are to believe them, this is the case with Yup'ik peoples. More generally, the lack of consideration among humans themselves breeds coarse attitudes towards non-humans. The 'natural' elements are confused. The glaciers get angry. The winds are no longer predictable. As for the pack ice, it breaks all the time or does not reform. Nor is the analysis in terms of morality unique to circumpolar regions whose inhabitants see their intergenerational ways of life disappearing along with the effects of climate change. It is found in the mountains of the southern hemisphere, right in the middle of the Peruvian Andes.

Among the Q'eros, one can communicate with the collective spirit of each living thing. Geremia Cometti quotes a shaman (*paqu*) who explains to the members of the village in which he lives that 'this stone has life from the moment it exists for you. If you see in her an *animu* [an essence that animates all beings], she will see the same in you.' Q'eros interact with non-human entities. No parallel world exists in their eyes. Spirits live alongside them and depend on the same world as them. Within this unique universe, interactions develop which do not exclude any being. They determine the balance of human society.

In his study of these people, the anthropologist shows that relationships with glacier spirits are the subject of rituals and offerings. In return, the deities protect the people. This is the contract. But the mountains are become bare. Icebergs are melting. They are upset: no one speaks to them any more. The shamans decry the fact that the rituals are disappearing and that they themselves have lost their influence. The ceremonies that brought together members of the community are no longer attended. Previously, the villagers were 'one person'. Today everyone thinks about their personal career and making money. People go to Cuzco and begin to believe in new religions. Few people are still in dialogue with the *apus*, the spirits of the mountains, of the ice or of rain. The shamans deplore this evolution of society. The community has forgotten that their children are 'sons of the earth' who were brought up 'like plants'.[38]

In this context, climate change becomes noticeable when the shaman can no longer 'lift the fog'. Indeed, 'rain, like snow, only listens to us if there is a collective will. In other words, if there is a collective will, they obey us, otherwise they do what they want.' The author of these remarks, the shaman Nicolas, is convinced that climate change 'is only the reaction of the earth and the *apus* vis-à-vis a collective loss of consciousness.' For him and others, Cometti tells us, the disappearance of rituals is one of the causes of the climate crisis. The massive departures of villagers have impoverished the spirituality of local communities. Impossible to get out of this infernal circle.[39]

Both kinds of arguments are maintained in many communities in the North as well as in the South. On the one hand, climate change is reducing the thickness of ice cover all over the world. This is an established, objective fact, and few doubt it today. On the other hand, glaciers and pack ice still make humans understand that they are not playing the right roles: selfish and neglectful of each other, they no longer manage to coexist with their living environments. By retreating, the glacial beings ask them to take a closer look at the consequences of their reciprocal attitudes. They invite them to resolve their contradictions.

Like sleepwalkers, humans have their eyes open, but their gaze is blank.

Respecting distance

The legitimisation of a system of thought is measured by its social structuring effects. Societies that define themselves in relation to natural entities integrate them into their collective ways of being; they transform them into partners. Kinship systems are never set in stone. Certain human attitudes make these relationships evolve, for good or for bad. By reacting to the singular behaviour of a few members of a given community, glaciers, icebergs or pack ice put the unity of the social whole to the test. This game between individualisation and cohesion is not a zero-sum game. Everyone has an interest in collaborating. You just have to observe the rules, one of which is the correct distance.

In most mythologies, too close a proximity between humans and non-humans is the source of endless trouble and harsh punishments. Claude Lévi-Strauss identified the trope, present in South America, of the 'canoe trip'. Two brothers embark on a canoe. One is the Moon, the other is the Sun. The first sits at the bow to propel the canoe, the second sits at the stern to stay on course. The Moon and the Sun must keep a 'good distance' between them, not move, otherwise they will fall into the water, and remain silent. The function of the myth is as follows: from the point of view of kinship relations within totemic clans, what is the correct gap which, for example, authorises human unions? Like the moon and the sun, a woman and a man will be neither 'too close' nor 'too far' – in this case, neither brother and sister nor male and female animal. In the first case, there would otherwise be incest. In the second, monsters would be produced.[40] Without the right distance, no social convention is worthy of the name. Without it, society cannot stand on its own feet. The scheme is valid for natural elements. The sun does not get too close to the Earth so as not to reduce it to ashes. Moreover, the right distance between the sun and the moon prevents night and day from lasting too long. Thanks to this salutary balance among the stars, salmon also periodically repopulate the rivers.

How does one determine, in societies, the distance that should be established between human and non-human beings? The answers differ depending on the case. The Tlingit guides who accompany John Muir on his excursions do not want to get too close to glaciers. Intrigued by his curiosity, one of them, Toyatte, declares with his hands on his hips: 'Muir must be a witch to seek knowledge in such a place as this and in such miserable weather.'

By going out to meet the glaciers, won't the naturalist disturb them? Unless he is a shaman, he risks disrespecting them. Since Muir does not have this quality, another guide, a Hoona, announces that there is no way he will touch the glaciers. He recommends that their group turn back under penalty of perishing, 'as many of his tribe had been, by the sudden rising of bergs from the bottom.'[41] If the guides had known that James Cook's sailors, off the Antarctic continent, were aiming their pickaxes at icebergs to extract blocks of fresh water, they would surely have judged such acts as perfectly reprehensible.

From the point of view of the natives, all these people are in contact with the glacial entities, but they are strangers to them. They walk on them and enter their caves freely. Sometimes they do not even hesitate to attack them. The glaciers and icebergs see them coming with their 'eyes like the moon'. They don't like to be stared at directly by strangers, let alone damaged. This is why, recalls Julie Cruikshank, people in the past tended to hide their appearance from the glaciers. They covered their faces and wore dark glasses. They hid their eyes.[42]

The freedom of movement provided by a smooth, firm surface is a 'gift' in the far North, as has already been explained. Travelling by sledge strengthens social bonds. But this freedom must be treated with 'great respect'.[43] The question of respect is crucial in Arctic societies that live with snow and ice. The rules of discretion are the same for all frozen masses, both in the mountains and at sea.

Paul-Émile Victor and Joëlle Rambert-Lamblin tell the story of an expedition of Greenland hunters with their families. A trip in sealskin or walrus canoes (*umiak*) beyond the limit of the ice is never a minor event. Everyone gets ready, as if anticipating a trip with a tragic outcome. No one is unaware of the cautionary protocols. They are repeated before leaving. Once seated on their knees in the boats, they paddle at a good distance along the Puisortoq glacier, whose ice spur gouges the sea. Their canoes move among icebergs adrift after falling into the water. They are afraid. They are even more afraid of the bergs that appear unexpectedly from the depths. Glaciers sometimes act like 'veritable underwater volcanoes'. From their 'basements' pieces break off and violently burst to the surface. These icebergs are invisible. No sound announces them. The threat seems unreal, so deep is it buried under the dark glassy surface.

When a canoe is broken in two by a mass suddenly thrown into the air, the hunters believe that the glacier is furious with the humans. They have passed too close to it or looked it in the eye. In retaliation, it destroys the *umiak*. The Igertewa glacier, located north of the Ammassalik region, is 'fierce' because it releases many massive icebergs from below. It forces hunters to paddle away from the coast, and all the travellers have to squint to see the shoreline on the horizon.

The glacier is so huge that it takes a good day to get around it. Before sailing towards it, the hunters and their families stop and make final preparations. They anticipate the moment when they will have to approach it:

Everyone eats to build up their strength and because, as long as the crossing lasts, they will not eat so as not to upset the spirit of the glacier. Then the crossing begins. In silence. Not a single word must be spoken; the oar strokes must also be silent, like the people. The children, frightened by everything they have been told, are huddled in the bottom of the *umiak* and hardly breathe. The mothers suckle their babies for fear that they will scream, which would inevitably anger the spirit of the glacier. When they reach about the middle of the glacier, the *umiak* stop and make an offering. At other times, valuable objects such as beads, pieces of iron or skins were thrown noiselessly overboard. During the entire crossing, no one except the helmsman, and only if he is expressly obliged to do so, should look at the front of the glacier.[44]

This description echoes the direct gaze taboo evoked by Julie Cruikshank in relation to the glaciers of Alaska. Wearing dark glasses is like not looking at the front of a Greenlandic ice tongue. Humans take these precautions because they calculate the risks of the environment through which they are travelling. They know that the masses are irritable and sometimes arbitrary. They are also aware that they have a normative function; they correct errors and rectify deviations in their behaviour. When the imperative of distance is not respected, glaciers and icebergs threaten to crush these boats made of skins.

The reasoning seems paradoxical. Ice induces a physical relationship, a sensitive body-to-body relationship.

The indigenous peoples live with it, in a relationship of permanent proximity. When they interact with glaciers or the sea ice, they never neglect the scale of things. But they sometimes believe they are walking in the clouds. No more background, no more surface, no more planes. Hunters experience the embodied reality of contact with their environment. All their senses are on the alert. Suddenly, there is an animal lurking under the snow cover, very close to them. The mountain glacier is the lair of a legendary bear or the hiding place of a huge snake.[45] The sea ice can turn over in seconds, as if a dragon were sleeping in the depths and would wake up with a start.

For those who think like this, there is no ambivalence. If the animal is outraged, it makes noises and gets bigger. Then it cracks like a thunderclap and changes shape. It retreats or abruptly descends a slope to crush an encampment. Along the ice floes, it gushes out of the deep water or falls from a glacial front to crush canoes. Shamans remind us that glaciers are like spiritual entities. There is something sacred about their faces. It is not appropriate to offend them by confronting their gaze without knowing them. This is the meaning of the rules that require humans, in the mountains and at sea, not to eat in their presence, to observe silence, to honour them with gifts and not to scrutinise them.

Here, it is the ice that governs.

The manners of indigenous people towards the powerful beings that inhabit their world have often been studied. David Abram reports on the speech of the Mattole Indians, who do not wish to disturb, let alone offend, the river that listens to their conversations. The Athapaskans of the north coast of California say that 'the water is watching you.' It has a 'precise' attitude. It is 'favourable or unfavourable'. Here, too, the rules are clearly stated, like rituals with no apparent room for negotiation:

Don't speak just before a wave breaks. Don't talk while crossing a current. Don't look at the water for a long time without looking away, unless you've been there ten times or more. Then the water in that place is used to you and is not bothered by your gaze. Older men can talk near the water because they have been in that place so much that the water knows them.[46]

If we read this passage carefully, the real issues to do with the taboo concerning the direct gaze become clear. The ban concerns first glances. It is not absolute. It requires more or less formalised approach rituals. The restrictive force of the taboo is reduced, or even erased, as soon as a relationship of familiarity is established. Regular attendance, a silent attitude, discreet gestures, all these criteria make it possible to inhabit a place gradually, as if one had been born there, and to obtain the assent of the people who live there. One can then undoubtedly look frankly at a glacier and its children, the icebergs, in the eye.

But the order of interactions is fragile. In a world extended to all living beings, there is a price to pay; it is that of a certain distance. Each relationship manifests this and demands that it be respected. Distance preserves the territories of other beings. It ritualises contact between the living. It makes it possible to set up meeting areas, to decipher the signs that are sent and not to encroach on the respective domains of each other.

In such a world, one does not behave with natural entities without following rules regarding food, noise and body postures. Deference is a way for humans to regulate their relationships with them and to establish cohabitation. The opposite behaviour, from indifference to rudeness, is a lack of consideration. Before acting, every human being must take into account the spirit of the entity that stands before him. He must decipher the marks of its interiority, interpret its feelings, thoughts and intentions. He must make himself transparent, remain circumspect and show wisdom in his attitudes. It is a mixture of standards and intuition.

In the wild, tact is necessary.

5

A Less Lonely World

Glaciers are the archives of the past, veritable open-air libraries. Their ice crystals hide infinite riches: layers of dust, gas bubbles, oxygen isotopes. Some ice-core samples in Antarctica show chemical signatures dating back more than 800,000 years. Samples taken out of the long metal tubes contain evidence of ancient events. The scientific community dates them on a long time-scale: a volcanic eruption occurred several millennia ago; the clouds had a certain temperature when certain snow-flakes fell and then crystallised. The crystals contain the remnants of the old atmosphere. The sky is in the ice.[1] This is why the gradual disappearance of the ice sheets is making mankind a little more amnesiac every day. We are losing our own memory. And the present itself is fading before our eyes.

An endangered species

In glaciology, any depth of snow is analysed in terms of 'mass balance'. The dynamics of accumulation and abla-tion of compacted crystals are compared. Variations in melting and calving phenomena are studied. It is clear that the gradual disappearance of glaciers is more than just a negative mass balance.

Gretel Ehrlich reminds us of this in a way that applies as much to the glaciers on the coasts of Greenland as to those in mountain ranges all over the world: 'Once ice loses its skin, texture, temperature, and shape, there's no getting it back. *La mer de glace* shatters, and the glacier's face, its toe and snout, and the storytelling tongue.'[2] The Mer de Glace, in the Chamonix Valley in France, is an animal dying in its den. The polar ice caps are also breaking up and diminishing. On land or at sea, all icy entities are in trouble. What affects them affects us. Their stories are our stories.

We have to push the argument to its conclusion: if glaciers are active and reactive beings, agents and not things, essential elements in the Earth's cycles as well as partners in social life, they are undoubtedly an 'endangered species' today.

This is a strong way of putting it. It would change the representational status of glaciers in Western thinking. The historian Mark Carey is working on the moment when the former fear of seasonal ice break-up turns into a passion the adventurous scaling of summits, completely reversing the polarities from anxiety to curiosity. The risks are no longer perceived in the same way. The emerging field of glaciology made this shift around the middle of the nineteenth century. Between 1830 and 1840, Louis Agassiz set up a rustic hut on the Unteraar glacier in Switzerland, in the Berne canton. He measured the movements of the giant to test his theory of moving glaciers. This scientific experiment was already encoding the values of physical discomfort and the heroism of heights. In the 1850s, the British Alpine Club was founded. The first guides to the peaks, passes and glaciers were printed. People found it interesting to walk on ice. Ice became 'recreational' just as it became a science.[3]

Generally speaking, institutions consider glaciers to be an 'endangered species' when they potentially endanger other living beings, including humans. Five major criteria are identified: a negative mass balance (the drastic reduction of volumes), scientific data to be saved (the chemical memory of the planet), an imminent collective threat (less fresh water per capita and eventually more polluted water), urgent protection of sites (in a legislative framework) and negative emotions (deprivation of an environment for relaxation and escape).[4] Our previous chapters are sufficient to show that two further criteria should be added: the foreseeable extinction of certain species living in circumpolar ecosystems and the fact that communities consider glaciers to be living beings in their own right.

The notion of danger is as much the result of a general observation on a global scale (all glacial areas are losing their mass) as of the anticipation of a state of affairs that would be even worse (the disappearance of these masses and the consequences on sea levels). Glaciers have been registered as UNESCO World Heritage sites. This was the case with the Sermeq Kujalleq in the Ilulissat fjord in Greenland. Since 2004, this place has been declared to be of 'exceptional value to humankind'. Together with others on the planet, it represents one of the 'irreplaceable sources of life and inspiration'. In this respect, it belongs 'to all the peoples of the world, regardless of the territory in which they live.'[5] The problem is that it continues to melt.

Carey asks what it is we are trying to save when we want to save glaciers in this way. Are we trying to validate the dream of an innocent nature that humans keep defiling? For whom do we want to protect these areas: for themselves, for the indigenous residents, for the world's tourists, for humanity, or for the Earth as a single huge organism?

The reduction of the sea ice and land glacial surfaces has sharpened the appetite of the economic powers whose project was to make new sea routes profitable and to exploit the wealth of land finally freed of its fixed covering. Mountainous heights and coastal fronts are home to the play of power and money. In the past, people feared the terrible approach of glaciers. Today, there is fear that they will disappear and the water tables will dry up. In the Peruvian Andes, there are increasing conflicts between local people and scientific and governmental experts over budgetary control of water resources and administrative risk management.

The market share of 'last chance' tourism is also increasing. This is the case for the Perito Moreno glacier in Argentina, whose dimensions remain gigantic. Going to see icebergs, knowing that they may be on their way out, is an ambiguous, even uncomfortable, pleasure. Spectators become eyewitnesses to a catastrophe unfolding before their eyes, as if they were standing on the threshold of a localised end of the world. Everyone then climbs back into the huge cruise ship, which contributes, through its high carbon footprint and fuel released into the water, to accelerating the rate of melting of the glacier it has just visited for photographic purposes.

If we stick to the spectacular aspect of disappearing glaciers, do we not validate a static, and somewhat biased, conception of nature? Why define the glacier as an 'endangered species' if it simultaneously becomes a 'neoliberal glacier'? Are national park enclaves reserved for a public of penultimate-day consumers?[6] This last point needs to be discussed on its own.

Empty or full?

Let's remember an expression John Muir used. The naturalist is witnessing the spectacle of glaciers calving icebergs. They create high waves as they drop. He records in his notebook that the noise breaks 'the calm dark nights when all beside is still'. The icebergs wake nature up from its comfortable bed. They wake her up with quite a fanfare.

This approach to wilderness has been subject to critique. Muir was considered to be the father of a kind of nature writing that turned ecosystems into stage sets. The writer would strip landscapes of their multiple life forms. He would reconfigure them according to his descriptions, constantly adapting what he sees according to his desires and to give his readers the impression that he is in the process of discovering something. He would focus on the 'process of seeing' rather than on the 'object seen'. Basically, he would be only a 'maker of views'. Hence it would follow that the silence in question would be only an artifice.[7] It is a harsh reproof. Muir was fascinated by the acoustics of glaciers and icebergs. There is no doubt in his mind: they are alive and well; they are wild animals! But one difference is highlighted. It is as if the desire for wilderness could not really be shared and that only a few travellers had the right to visit it. What conception of nature does this approach reveal?

The Wilderness Act was signed into law during a session of the US Congress in 1964. This administrative text sets in stone the intellectual objectives of national park policy in the United States. The general principle is that wilderness is defined in contrast to cities and other built-up areas. It is a natural area, 'an area where the earth and its community of life are untrammelled by man, where man himself is a visitor who does not remain.' Three general criteria are taken into account. They are supposed to be strictly adhered to. The space in question '(1) generally appears to have been affected primarily by the forces of nature, with the imprint of man's work substantially unnoticeable; (2) has outstanding opportunities for solitude or a primitive and unconfined type of recreation; (3) has at least five thousand acres of land or is of sufficient size as to make practicable its preservation and use in an unimpaired condition.'[8]

Purity, solitude, immensity – these are the conditions. In other words, the wilderness must resemble the beginnings of a barely inhabited Earth. It will be a virgin space that excludes all society. It will play the exceptional role of a large escape zone, admitting only a few humans.

This excerpt from the 1964 act is well known. The historian William Cronon spoke out against such a conception of nature. His view is that the wilderness is 'quite profoundly a human creation'. It nurtures the dream of an untouched and unpopulated nature. By embodying the fiction of a 'pristine sanctuary', it passes judgement on the state of society. After many years of consultation at the highest political level, it was decided that parts of the planet should be eternally free of the 'the contaminating of civilisation'. This decision was taken at a time when the romantic theories of the sublime and the imaginary of the 'Frontier' were spreading in the United States. Suddenly, these wide open spaces provided an image of a new horizon of possibilities, away from the immorality of the city.[9]

Another historian, Frederick Jackson Turner, dubbed the westward advance of European immigrants and eastern residents the 'frontier'. During this episode of American history, the nation had expanded and founded itself as a nation by confronting a hostile nature. It is well known, in this history, that the Indians lost the places in which they lived. They were expropriated from their territories. Between 1850 and 1900, the American plains were domesticated, emptied of their buffalo, wolves and humans, all three massively exterminated. Settlers moved onto the Great Plains without cohabiting with the beings that inhabited them. Wilderness 'functions as a reification of the wild'.[10]

Turner stated that the 'Frontier' phenomenon did not persist beyond the 1890s. Cronon argues that the policy of national parks in the United States merely prolongs this 'myth of origin' of conquest of the Great Plains as 'the nation's most sacred'. First, it brought together men who were fleeing the soft communities of the cities. It valued 'wilderness experience' on the trails of untouched nature. Then the elites stripped the vast spaces of their virile connotations and turned them into places of 'recreation'. Silent, peaceful valleys replaced images of wildness. City dwellers flocked to relax and indulge in their reveries. The wilderness was thus, from start to finish, a 'siren song of escape'.[11] This misunderstanding made it impossible to reinvent the place of humans in nature.

First Nations representatives share this criticism. One example is the Tatshenshini-Alsek Wilderness Provincial Park in the glaciated areas of Mount St Elias in Alaska, which was renamed Tatshenshini-Alsek Provincial Park in 2000.[12] The concept of wilderness has disappeared. The message is clear, this notion carries only one history. It prohibits the recognition of other ways of conceiving of the wild.

Among these other approaches, that of Henry D. Thoreau is central. A famous phrase summarises it: 'In wildness is the preservation of the world.' This phrase is too often misinterpreted. The terms are confused. Many commentators read 'wilderness' instead of 'wildness'. They equate 'wilderness' with 'wild life'. They pretend that Thoreau and Muir shared the grand narrative of unspoiled plains, forests and mountains. Few of them also note that this call is about human society before it is about ecosystems.[13]

Let us recall the three main meanings of the notion of wildness.

Thoreau first addresses his fellow countrymen. He notes their state of servitude and begs them to turn to the vitality that lies within them. He asks them to emancipate themselves from the moral and material constraints that shackle them and to 'simplify' their existence at all costs. Otherwise, they will never succeed in living on the essential alone. To be 'wild' for a human is to refuse the patterns of alienation imposed on him by his fellow human beings. With the sole aim of being entirely on one's own. Wildness is a force for freedom. It is the name of the inexhaustible active principle that provides each person with their autonomy.

Wildness is nonetheless a shared affair, a goal of life in common. Thoreau is convinced that the more autonomous individuals are, the more collective bonds are forged and strengthened. The freedom of some increases the freedom of others. Obeying one's own laws does not prevent one from conforming to social norms. But they must be fair. If they prove to be unjust, the individual is all the more entitled to challenge them as he or she draws energy from his or her acquired freedom. Thus 'resistance to civil government' (posthumously renamed 'civil disobedience') contributes to social cohesion. In two words: wildness is a political 'tonic'.

Thoreau's formula concludes with universal value; wildness is the common trait of all living beings, human and non-human. His botanical notes are a eulogy to the ardour of floristic biomes. He details the vigorous pistils, stems and petals that line the forest paths. He weeds among the beans that grow in his Walden field at a dizzying pace. The pages of his diary are full of similar observations. And when he reports, on 7 May 1852, that he cannot find the words in his language to respond to the modulations of birdsong, he goes to great lengths to describe this musical aspect of 'wild life'. For Thoreau, these three senses complement each other. The living beings who synthesise them in themselves are fully 'wild'.

Conceptions of 'wilderness' and 'wildness' come up regularly in academic and public debate.[14] It is difficult to reconcile them. Where do glaciers and icebergs fit

into these discussions? Glacial areas have long been perceived as non-humanised areas. An aesthetic value was
attributed to them very early on – that of the sublime,
which we mentioned at the beginning. In this spirit,
European explorers measured themselves against the
frozen masses. They tested their patriotism and their
particular sense of adventure. Many of them experienced the ice expeditions as journeys to a final 'Frontier'
after the 'Frontier'.

Icebergs are nowadays the icons of the poles. They
are monuments of an attractive fragility. Contemporary
travellers flock to both ends of the inhabited world to
contemplate these emblems of extraordinary nature.
After first haunting explorers, the solitude of icebergs
has become an object of recreation. Icebergs embody a
new wilderness accessible to wealthy city dwellers of the
globalised metropolis. People come here mostly as 'visitors who don't remain'. They see impressive spectacles.
Then they return home with mixed feelings, melancholy
and/or exhilaration.

Why are icebergs fixed in a story of isolation and
empty nature? One can feel close to a colossus drifting
on the waves. But its glorious solitude is an illusion.
Icebergs are never alone. They carry biotic environments. Myriad auxiliary organisms of various sizes
hang on to their sides and, thanks to them, can perpetuate their life cycles. Along with glaciers, they are both
respected and feared entities in human societies. We
have seen that they play critical roles in revising social
rules and calling for order. They even help to resolve
moral dilemmas.

To abandon the representation of the solitary iceberg
is to begin to repopulate the ice worlds, to 'rewild [their]
wilderness'.[15] It is time to rediscover wildness wherever
it is found, along small forest paths as well as on the
Great Plains and on the walls of icebergs. It is the wild
that is disappearing.

Suddenly, nothing at all

In the Inuktitut language, *auyuittuq* means 'the land that never melts'. It is the name given to glaciers.[16] Glaciers embody the deep time of the Earth, the memory of its elemental phases, the memories of the ancestors who lived with them. They are a link between the past and the present, like a promise of continuity. Their antiquity is part of the practical and mental life of circumpolar human communities.

Nowadays, no glacier can bear this name because of the general disappearance of the frozen expanses. What sense would there be in defining a glacier as 'the land that melts'?

In the summer of 2014, glaciologist Oddur Sigurðsson reached the top of a volcanic prominence in Iceland after several hours of walking. He wanted to check whether the film of crystals covering the 'Ok glacier' (Okjökull) still allowed him to talk about a glacier. He 'suspects' that the thin snowfalls will prove the contrary since the glacier has not been renewed for some years. On his return from his expedition, he tells the scientific community that the Okjökull must be reclassified. It has turned into 'dead ice' as a result of global warming.

Five years later, at the initiative of anthropologists Cymene Howe and Dominic Boyer, a group of about one hundred people repeated the glaciologist's journey. It was a group made up of personalities from the worlds of politics, the media, science and the arts. Its aim was not scientific this time. It was a pilgrimage of mourning. It was about commemorating the loss of a person. The writer Andri Snær Magnason composed a bilingual text in Icelandic and English. The words are engraved on a copper plate (see figure 6). The following can be read by anyone completing the journey despite the fog or seasonal snowfall: 'A letter to the future. Ok is the first Icelandic glacier to lose its status as a glacier. In the next

Figure 6 Bréf til framtíðarinnar / A letter to the future.
Photo © Cymene Howe.

two hundred years, all our glaciers are expected to follow
the same path. This monument is to acknowledge that we
know what is happening and what needs to be done. Only
you know if we did it. August 2019, 415 ppm CO_2.[17]

During the ceremony, the death certificate was read
out by Sigurðsson: 'Death by heat. Death by humans.'
The memorial stone is solemnly placed on the ground
by children and then fixed with its plaque to the spot
at 64° 35' 29,88" north latitude and 20° 52' 15,18" west
longitude. One can imagine the scene: dignified attitudes, silence is prescribed, only the wind's voice can be
heard whistling.

By agreeing to write a text and to participate in this
unprecedented memorial expedition, Magnason wanted
to invite each of us to connect with the 'intimate time
of the future'. Children were part of the journey for at
least two reasons. On the one hand, solidarity between

generations is indispensable; on the other hand, they will judge the situation later on. In order to help them enlighten their minds at that time, the writer wanted to note down, and date, the concentration of carbon dioxide in the atmosphere.

It is well known that commemoration ceremonies have a dual purpose, in the short and the long term. They do not only recall events that have traumatised societies. They also pay tribute to the victims who have passed away by raising painful questions. Ultimately, they aim at a return to a so-called normal existence and shared agreements. One remembers in order to succeed, one day, in mitigating what hurts so much, or even to stop thinking about it. Forgetting is part of memory, that is the assumption. When the period of mourning seems over, when memories fade and direct witnesses disappear, many ceremonies are abandoned.

The pilgrimage to the top of the glacier shares some of the features of a classic post-trauma commemoration. But the victims' relatives are not present to listen to the official speeches and attend the ceremony. The participants did not lay a plaque in order to go on with their ordinary lives as best they can. The purpose of their open-air assembly is different: to witness the passage of a glacier from life to death and to refuse to let other glaciers die out in indifference. This is a fight against anonymity and for a form of life to be recognised. It is the inauguration of a process of ritualisation that is, in principle, unlimited and generalisable. It is the inscription, in the ground, of their claim to remembrance, and they want everyone to respond. There is no reason to end the work of mourning. The drama of this disappearance must not be forgotten. Unlike a classical commemoration, the ritual does not aim to pacify the emotion of the loss. It is a call to continuous action. The awakening of consciousness must last. A sword of Damocles hangs over our heads.

'If we say something has died, can we also say it once lived?' asks writer Lacy M. Johnson. Familiar with glaciers and volcanic regions since he was a child, Oddur Sigurðsson tells his interviewer, his eyes a little misty, that 'a good friend has left us'. A glacier means a lot to him. The ice crystals contain pollen carried by the clouds, dust from volcanic eruptions, and even traces of wars between humans. They incorporate the history of the Earth as well as that of humanity.

But, above all, a dying glacier makes it clear that an internal dynamic becomes inescapable once it has begun. When the ablation of ice crystals prevails over their accumulation, the melting of the mass accelerates with no possible return. Geophysicist Marco Tedesco calls this process 'melting cannibalism': '. . . it's melting that's feeding itself. . . . Rising temperatures are promoting more melting, and that melting is reducing albedo, which in turn is increasing melting.'[18] And so on.

The researcher is not the only one who feels empathy for the receding glaciers. Many Icelanders use words in this situation that describe the psychological life of humans. They tell glaciologist and geographer M Jackson that a glacier 'watches' them and 'feels' emotions. It 'interacts' with them, blossoms like a flower and 'breathes' like any living thing. Many are even convinced that it has a 'cognitive awareness'. The glacier is also capable of being 'happy'. It often has 'not enough to eat' and its nights are filled with fears and dreams. Finally, one can tell when it is 'healthy, not healthy'. More than ever, the melting of the glaciers reveals a life intensely lived.[19]

It is enough to visit a glacier and associate memories with it to feel close to it. Breiðamerkurjökull, in the southeastern part of Iceland, is said to 'bloom' in good weather. It opens up and shines with the other glaciers in the area. In an interview with M Jackson, a resident called Dögg says that you can even 'see what they are thinking. In the spring they turn white, they hide. I don't like them in summer because they are black, they are dirty and hidden and they cannot speak. This is life. But then, in autumn we have the rains here, and they are bathed and washed, and the bad skin is sanded away, and then they are smooth and clear and ready for snow dressing. This is the glacier life.'[20]

Anthropomorphism is here carried to its clearest expression. The common objection to such descriptions is not much help. Firstly, the anthropomorphisation of glaciers is not intended to scientifically demonstrate their nature as living beings. The psychologising vocabulary simply attests to a long history of intimate links, sometimes visible, sometimes invisible. Secondly, animistic societies, and those that are not animistic but possess certain traits, describe relations between non-humans as much as relations between humans and non-humans in terms of interhuman relations in general. The 'sociocentrism' of people who live near glaciers is their only language tool.[21] It does not impose any representation of the world. It reflects a central aspect of their daily life with beings made of ice.

The Ok pilgrimage received a lot of media attention. Perhaps, some say, because it took place in a land forged by glaciers, which one had hoped would be safe from decisive human impact. Perhaps, others say, because it inspired the desire to record the names of ignored glaciers and to baptise all those that still lack them. This desire to name does not betray any will to extort or arrogate territory. On the contrary, in other parts of the world where glaciers are under threat, local people are invited to suggest names so that none of them disappear incognito, without a burial ceremony. This is a way of keeping the dead close to us and continuing to care for them.

A dying glacier no longer makes sounds. It no longer cracks the joints of its bones or releases its last icebergs into high lakes or coastal bays. Cymene Howe recounts the dismay of an Icelandic friend who no longer hears the 'screeching sound' of ice on water, the clamour of its banging against the hulls of ships docked at her childhood port. These sounds frightened her as a child. Today, she realises that she 'misses' them. The anthropologist discerns in this memory much more than a sensitive experience that is missing. The silence of the coastline now sounds like a 'requiem in the future subjunctive'. With some brutality, the 'sonic disintegration' of the northern regions is 'silencing' the sea ice. It disturbs the humans themselves. This kind of phenomenon forces us to formulate more precisely the affective relations we have with melting ice.[22]

Proof through emotion

Between 2018 and 2019, a group of scientists from the universities of Greenland and Copenhagen conducted a survey in six inhabited areas of Greenland. The study of the data collected reveals some major findings.

Most Greenlanders do not believe that they will benefit from the melting ice. They see the melting ice ruining their way of life and they know that it is destroying ecosystems. For them, ice is still a necessity, as are dog sledding and halibut fishing in the holes in the sea ice. But the escalating amount of rain is now the ice's worst enemy.

Top of the priority list of concerns is the increasingly unpredictable climate. Then come the loss or thinning of the sea ice, the disappearance of permafrost, heavy storms, melting glaciers, polar bears roaming near villages, and the extinction of certain animals and plants. In the percentages, the differences are small. It is obvious that these factors are closely related to one another. Finally, the social anxiety and 'post-traumatic stress' caused by the impacts of the climate crisis on these territories are clearly perceptible. There is a general feeling of powerlessness. This concerns above all food security; hunting on the sea ice is no longer practised, and imported products are excessively expensive, not to mention their quality. As for sled dogs, they are no longer used.[23]

Dr Courtney Howard was president of the Canadian Association of Physicians for the Environment. She and other residents interviewed confirmed the study's findings that the gradual disappearance of the sea ice is isolating people from hunting, travelling and socialising. The ice connected the coastal villages in the absence of roads across the glaciers and fjords. It was the 'umbilical cord' that bound friends and families together. Greenlanders are increasingly living on 'islands', as if in an 'archipelago'. The sea is even becoming stronger than the ice, and more dangerous too. Many children do not know how to swim. If they were to fall accidentally into the water, they would drown. In general, Greenlanders are in mourning for their life forms. For them, the sea ice is everything.

Courtney Howard believes that the word 'solastalgia' aptly describes the emotion felt by people experiencing the effects of climate change in the Arctic.[24] The term was coined by Glenn Albrecht some fifteen years ago. The philosopher combined the word 'nostalgia' with the two Latin roots of *solari*, for 'consolation', and *desolare*, an infinitive verb meaning 'to depopulate', 'to ravage', 'to abandon'. What exactly does this neologism mean?

'Solastalgia' is a term used to describe the pain that results from an 'ongoing loss of solace', a kind of endless worry. The rapid disappearance of the familiar environment gives rise to a strong feeling of abandonment. Albrecht wants to name the kind of emotions that many people feel deep down. He goes on to say that this is 'a chronic condition, tied to the gradual

erosion of identity created by the sense of belonging
to a particular loved place and a feeling of distress, or
psychological desolation, about its unwanted transformation. In direct contrast to the dislocated spatial
dimensions of traditionally defined nostalgia, solastalgia
is the homesickness you have when you are still located
within your home environment.'[25]

Albrecht himself witnessed the methodical destruc-
tion of the Upper Hunter Valley in Australia by the
mining industry. He knows what it means to ruin an
ecosystem. He is interested in the psychological effects
of the changes that climate change is imposing on natural environments. He warns from the outset against any
confusion; solastalgia is not a mental illness. It is not
diagnosed in the same way as a doctor would diagnose a
patient suffering from severe depression. It is an anguish
that reveals and signals a psychological state as much
as a state of the external world. Solastalgia is an echo
chamber. It expresses the existential disarray of those
who no longer recognise their land. Albrecht notes that
the First Peoples of Vancouver Island, British Columbia,
used his term when they spoke in the media about the
gradual erasure of the Comox glacier.[26]

Solastalgia is thus the name for the helplessness felt by
residents of territories marked by man-made disasters.
The concept specifies the affective context of disorientation that follows the loss of an 'endemic sense of place'
due to the damage (fires, hurricanes, drought, floods,
melting ice) caused by the extractive industry, pollution
of the soil, atmosphere, oceans, and global warming.
Confronted with changes in their living environments,
indigenous populations are finding that the landscapes
they loved have become 'unrecognisable'. The land has
been degraded, the outlook devastated. In the past, a
hill was the memory of a community. Now, when it still
exists, it looks hostile. Places die with their visible and
invisible geography.

Barry Lopez describes the ties that bind indigenous peoples to their lands in the far North. The breaking of these ties is the source of infinite harm. Territory is not just about visibility. It is its *'invisible'* part that contains collective stories, personal memories and intimate dreams. A place only seems empty if one has not lived there for many years. Otherwise it is rich in signs. Those who have a long relationship with the nature around and within them suffer when it is brutalised. The reason is simple: they are connected to their environment 'by luminous fibers . . . to cut these fibers causes not only pain but a sense of dislocation.'[27] It is like amputating the four limbs from a body and extinguishing the glow of memory; they no longer know who they are.

The writer calls this invisible part of the relationship that the Inuit have with their environment 'spirit country'. An individual's identity is not limited to his or her body. The maps of this country are not quadrangular. They do not detail a geodetic expanse. They recapitulate, sometimes on the ice itself, more often in their heads, a territory known only to them and which others cannot perceive with the naked eye. Generally linked to hunting, to a birth or to a mechanical accident, they are faithful to the relief; no creek is missing, no mountain is forgotten. They indicate places, not empty spaces. They contain stories that feed the imagination and structure practical life.

In the far North, the ice cover has been reduced considerably in recent years. Hunting grounds among grounded or ice-bound icebergs have disappeared. Many 'fibres' have been cut. The link with dreams and memories encoded in these places is broken. The ice of many years no longer overlaps. It rarely forms the ridges that used to be tens of metres high. The pack loses its relief. There is less diversity of ice. The erasure of sensitive landmarks reinforces the feeling of being uprooted without experiencing any physical exile.

Jacopie Panipak, an elder from Clyde River in Canada's Nunavut, mentions his homesickness when he can no longer see the contrasting landscape of 'sea ice' stretching to the horizon. He experiences exactly the feelings described by Glenn Albrecht and Barry Lopez – the solastalgia of the ties that bind to the living environment. He guesses that the melting of the ice no longer depends on the regular play of the seasons. He is not the only one. No one is fooled. The members of the community know that the disappearance of the sea ice is not a simple disappearance. They interpret it as a loss in the strongest sense. The ice is not disappearing only to re-form the following year. It fades away physically, like a person who dies. Like hunters who leave in search of food and never return to camp, it reappears only in dreams.

In the Inuktitut language, it is not in the nature of the glacier to melt permanently. It shrinks only to grow again. It maintains and reproduces itself. The ice cover of an area may thin out in the summer. The so-called old snow mass, piled up over centuries, is resistant to seasonal variations. Glaciers behave like animals that build up fat for the winter and become thinner during the long summer days. But they are no longer renewed today. The snows are no longer eternal.

The resistance of rocks

All climate analyses lead to one definitive conclusion: the dynamics of the melting of the ice caps have accelerated dramatically in recent decades. The general level of the oceans is likely to rise without delay. The world's coastal areas are under threat. Aka Niviâna and Kathy Jetñil-Kijiner are deeply implicated. They react to this urgency by writing a prose poem, performing its cadences in a short video (see figure 7).

Figure 7 Rise: From One Island to Another, a film by
Dan Lin, Nick Stone, Rob Lau and Oz Go, produced
by Bill McKibben and 350.org
(https://350.org/rise-from-one-island-to-another/)

The text features two young women exchanging gifts.
In the video, the authors embody these characters and
appear on screen alternately. The story is as follows: one
of the women lives in Kalaallit Nunaat in Greenland. She
welcomes the other woman who lives in the Marshall
Islands. The first carries stones from Nuuk, the second
brings shells from Bikini Atoll and Runit Dome. The
latter two places were used as testing grounds for a
number of US atomic tests in the immediate post-war
period. They still contain radioactive waste.

The short video unfolds like a gift exchange ritual.
The 'sister of ice and snow' and the 'sister of ocean and
sand' greet each other. A sisterly relationship is established. In keeping with the ceremonial of hospitality,
they refer to each other by their geographical origin and
describe the lands of their distant ancestors. Then they
join hands and extend them, holding the material gifts
whose spiritual value they explain.

The shells and stones, traces of the first islands, illustrate local legends passed down from generation to
generation. In the Pacific, it is the legend of two sisters,
one of whom decided to stay by the side of the other
who had magically turned to stone. In Greenland, it
is the legend of Sassuma Arnaa, the 'guardian' of the
oceans who lives in the depths. Whales, rivers and icebergs are the 'children' of the long-haired primordial

'mother'. She observes humans with patience and indulgence. Today, she frowns and worries because she sees the hold of evil passions, 'greed in our hearts / disrespect in our eyes' that endangers her offspring. She gets angry

when we [her children] disrespect them
she gives us what we deserve,
a lesson in respect.
Do we deserve the melting ice
the hungry polar bears coming to our islands
or the colossal icebergs hitting these waters with rage?
Do we deserve
their mother
coming for our homes/for our lives?

Those present offer 'testaments' of a world that is going away. When one of the women tells of the rising oceans that will submerge her archipelago, the other tells of the glaciers that are 'groaning / with the weight of the world's heat.' From the middle of the film onwards, a long, haunting note accompanies the poem spoken aloud until the end. It is the cry of the glaciers echoed by the drifting icebergs. Wars and nuclear tests have produced waste that ends up not only in the waters of the Pacific but also in the glaciers of Greenland. Today, the ice is disappearing. This is followed by a silent shot, longer than the others, over a stretch of bare rock.

Climate change is blamed on that part of humanity that destroys living environments to the point of preventing them from renewing themselves and turns minds to money. Glaciers are roaring because 'colonising monsters' are massacring ecosystems with their gigantic machines. As a result, humans do not respect the right distance that connects beings to one another, the distance that allows each one to honour 'life in all forms'. The imperative of the right distance is not observed. The multiplication of raging icebergs is a bad sign. To try to stop this process, which has been under way for several decades, we must 'rise'. There is only one solution to the rising waters and the threat of sinking into the ocean: 'choose stone', to petrify oneself in the image of the legendary sister, to cling to one's place like a shell to its reef.

The film stages a new genre of *potlatch*. In the analyses of Franz Boas (on the Indian societies of the American Northwest) or Bronisław Malinowski (on the populations of the Trobriand Islands of Papua New Guinea), *potlatch* refers to a festive phenomenon of gift and counter-gift of material objects and moral values. Tribal leaders know that, by giving and giving back, they are exchanging respect and creating ties of mutual obligation. Rivalries are still strong. *Potlatch* is a trade in which one's superiority is tested.

The two young women's reciprocal gift-giving, in this case, is an exchange of equal value. It attests to their solidarity and their desire to preserve the world by acting in their own places. To prefer anchoring to wandering, becoming a stone to endless exile, is to be responsible. This is their freedom. This is their wildness, to use Thoreau's word.

The Ok glacier in Iceland was already dead when glaciologist Oddur Sigurðsson visited it in 2014. The commemorative copper plaque placed at its site five years later is an intensely solastalgic object. It is a reminder of the importance of our connections with distant environments and the need to preserve them. Disaster is similarly at work in Greenland and the Marshall Islands. Aka Niviâna and Kathy Jetñil-Kijiner want us to react, to feel that we are all islanders. They denounce the lack of concrete measures to stop the melting of the Arctic glaciers, which is gradually raising the level of the oceans and threatening to swallow up many Pacific islands. They argue that this is not just an 'inconvenient truth' for a few, it is now everybody's business.[28] No place is isolated any more. Local realities are now inseparable from each other. Islands are no longer cut off from the world. The effects of climate change in Greenland are being felt in the Marshall Islands. This interdependence proves that the world has no natural borders.

The call for responsible action aims to end the era of silence on the real consequences of the technical and industrial developments that is hitting the extremities of the world with full force. It is what the philosopher Donna Haraway calls 'response-ability'. Aka Niviâna and Kathy Jetñil-Kijiner rely on the inexhaustible 'capacity' of humans and non-humans to associate and 'respond' to each other.[29] They rely on the desire to inaugurate a dialogue that will not exclude any being. In the end, they have created a cosmopolitical narrative that corresponds to the diagnosis of the disappearance of ice. It is indeed a question of creating an awareness of the world from its damaged places.

In the film, the raging iceberg and the groaning glacier are not entities superfluous to human society, nor are they incidental details in some grand panorama. They are beings included in relationships with other beings, genuine co-agents. They should be cared for in at least two ways: as their own biodiverse environments and as members of 'collectives' anchored in territories that are threatened with extinction. The representatives of the equatorial community of Sarayaku present at COP 21 in Paris in 2015 reminded us of the urgency of broadening our scope to include living things in general. They stated their conviction that the future of the planet lay in preserving 'the material and spiritual relationship that indigenous peoples have with other beings who inhabit the living forest.'[30] This was a way of telling nations that entities to which they do not attribute any life are nonetheless 'alive' and that we are closely dependent on them.

But we have seen that these relationships take various forms. The 'social plasticity' of glaciers and icebergs is enormous.[31] It can be tested with multiple approaches. The task of finding a common language is not easy.

What separates and what unites

Here we come to the point where a difficult question is formulated: how do we reconcile the stories in which glaciers react to human attitudes by reducing their mass, or overflowing their beds, with the scientific equations that stabilise mechanical laws of ablation and fracturing? How can concern for the Earth include the beings that other humans consider to be alive, with their respective conceptions, and remain true to the spirit that tests, verifies and corrects the data of experience? In other words, is 'an animistic openness to the world the enemy of science'? Anthropologist Tim Ingold provides a decisive answer: 'Certainly not.'[32]

The problem is not just theoretical. It affects everyday life because of climate change being on the rise. Everyone is conducting experiments in their own localities. Scientists study glaciers. They patiently probe and measure them. They weight the calving of icebergs and model their drift trajectories in multiple variables. They record their voices and scrutinise the life forms that always accompany them. Local histories are available that describe them as beings that punctually regulate the deviant course of societies and restore order to the perceived world. These masses are not ornaments in some inert setting for those who live there. Language and practices bear witness to this. Many day-trippers see them, eventually, as the last relics of an extraordinary nature that is disappearing. They take advantage of an exceptional moment in their existence to celebrate their spent beauty.

Icebergs are sometimes places where one imagines oneself alone in the kingdom of the sublime, sometimes as underwater environments where unicellular micro-algae proliferate and provide a large part of the world's oxygen. Glaciologists refine their tools and interpret the history of the Earth by surveying spaces that local residents consider to be animistic zones of interaction. For some, icebergs and glaciers are an intense visual and aural experience. For others, they carry the past and future of the atmosphere or the ocean. For still others, they help a community to survive. All these multiple meanings are suddenly brought to life when a glacier calves or an iceberg turns on itself.

Scientists who have spent years in the field are nevertheless convinced that glaciers and icebergs have personalities. Indigenous people mourn the rapid melting of these 'critical non-humans' with whom they had centuries-old social relationships.[33] The episodic tourists, meanwhile, are increasingly attentive to the stories of the original peoples and the ecological role of frozen masses. Sometimes they cross paths with each other. There is no opportunity to discuss their respective ways of seeing. Each one nevertheless feels a deep empathy for glaciers and icebergs. It is as if a minimal animism, spontaneous and devoid of any systemic framework, potentially unites the experiences.

The significance, therefore, of moral arguments that invite traditional rituals with non-human entities into discussions about the rate of anthropogenic impact on the planet's climate cannot be overlooked. Geremia Cometti rightly reminds us that, in our global context, 'only an inter-ontological analysis, taking into account indigenous cosmologies and their explanations of the phenomenon and their practices, can give us a more complete picture of reality.'[34]

For their part, Catherine and Raphaël Larrère stress that 'dualism cannot be escaped through monism.'[35] It would be a mistake to believe that animism can repair the mistakes made by the overly productivist Moderns or that it is easy to ask everyone to abandon, for example, the concept of 'nature'. Generally speaking, no one changes their representations of the world by snapping their fingers. To put it another way, it is as difficult to suggest to a non-animist as it is to an animist that they change the categories they use to decipher the environment in which they have evolved.

There is an option. It consists of shifting the grids of interpretation by studying the points and lines where they meet. To do this, we must listen to different voices and not consider them as superstitions from another age, as false beliefs. These voices are evidence of an 'intelligence that could inform science'. Julie Cruikshank observed in 2005 that local views about glaciers are still 'relegated to the domain of "culture", thoroughly distinct from scientific studies of "nature".' She deplores this because such compartmentalisation of academic disciplines prevents the invention of appropriate ways of 'interweaving' narratives and methods.[36] Today, many people concerned with these issues feel that the disappearance of ice should prompt us to try harder to find the 'vocabulary that captures the subtle changes in peoples' engagement with their environment.' The reason is simple: the retreat of the masses disrupts ways of living 'here and now'.[37] This applies to both human and non-human beings.

The stakes are high.

The less ice there is on the Earth's surface, the closer we are to that moment that Edward O. Wilson called the 'age of loneliness'. To support his expression, the biologist has coined another term, 'Eremocene', from the Greek *eremos*, meaning 'desert'. Both refer to the same future state of affairs: an age in which only humans, cultivated land, 'as far as our eyes can see', and domesticated animals would remain.[38] This age would represent the logical continuation of the Anthropocene and the ongoing sixth mass extinction of species. It would be the new symbol of a planet on which no wild life persists. It would be the triumph of a humanity contemplating itself in the distorted mirror of its work. The silence of nature would be the proof that ecosystems, the diversity of their populations and habitats, would be exhausted.

The natural snow cycle is now diminished. Snow no longer has time to freeze deep enough to rebuild these gigantic masses, which fracture normally under the effect of internal pressures. The glaciers do not taper off and then regain their volume. They look more and more like their children, the icebergs; they are bound to disappear quickly. If all the sounds fade away, if snowflakes no longer pile up on the rocky bases, if no ice covers the Earth, the world itself becomes lonely.

How can we escape this 'age of loneliness'?

Glaciers and icebergs make it clear that we need to abandon the proprietor impulse. Just after the Second World War, Aldo Leopold recommended (posthumously) a shift from using the Earth as a 'commodity belonging to us' to a 'community to which we belong'. In his *Sand County Almanac*, the ecologist continues his analysis by stating that 'there is no other way for land to survive the impact of mechanized man.' The abandonment of the owner morality must be accompanied by a significant 'extension of ethics'. Everyone must succeed in producing in themselves 'a mental image of the Earth as a biotic mechanism [because] we can be ethical only in relation to something we can see, feel, understand, love, or otherwise have faith in.'

These words are transparent; there is no purely normative ethics. The planet is an organic whole. It requires a 'social consciousness' that does not forget the work of the senses and that integrates other living beings. The ethic in question requires that the 'boundaries' of this new community be extended not only to the soil, water, plants and animals but also, let us be careful to add, to glaciers and icebergs. In Leopold's view, *Homo sapiens* would do well to establish a more sensitive relationship with all beings if he is ever to become a true citizen of the Earth.[39] He needs to love the world in order to safeguard the abundance of wild life.

No one is a prophet in this field. We may feel that we are steering the ship. But we are not the voyage's purpose. Humanity is a species among other species. It is part of the course of evolution. Like other groups of living beings, it does not always understand the effects of its own actions. Nor does it understand in advance the real purposes of its attitudes. But we can know, like Socrates and Montaigne, that we do not know much and continue our work of interpretation in a condition of ignorance. This humility would imply that we would no longer be interested only in human roles. It would allow us to rewild ourselves (in the sense of wildness) to become familiar with other living beings and, perhaps, to dispel the illusions of any species egoism.

In the face of a floating tabular iceberg, a shoreless ice floe, a steep coastal glacier front, no conquering human pronoun is worth a jot. Glacial masses and expanses render many of our 'self-centred narratives' obsolete. They develop forms of life where we think everything is dead and force us to see the world 'from the outside' (to use Val Plumwood's phrase). They bring us back to our status as members of the same earth community. Rewilding is not only a method of regenerating biodiversity, a way of letting habitats renew themselves. It is also another name for discretely writing humans back into ecosystems.

Like a common language for us all, and a remedy for the solitude of the world.

6

Thinking Like an Iceberg

In one of the chapters of his *Almanac*, Aldo Leopold addresses cowherds on mountain pastures who are afraid of the wolf. He recounts the moment when his life changed. One day, as a hunter, he shoots at a she-wolf and her pack of cubs as they cross a river ford. When he sees the 'green fire' of life disappear in the mother's eyes, he understands the mystery that unites not the hunter and his environment but the wolf and the mountain. He guesses that the wolf regulates the possible supernumerary deer and helps to prevent trees from becoming too defoliated. If it wants to maintain the balance of life in its biome, the mountain needs the wolf to avoid the fear of too many deer. The farmers would benefit from extending the relationship they have with their cows to the environment that includes all beings, together with predators. They would hear the cry of the wolf with a different ear:

> A deep chesty bawl echoes from rimrock to rimrock, rolls down the mountain and fades into the far blackness of the night. It is an outburst of wild defiant sorrow, and of contempt for all the adversities of the world. Every living thing (and perhaps many dead one too) pays heed to that call . . . Only the mountain has lived long enough to listen objectively to the howl of a wolf.

The deer is frightened, the pine anticipates the blood on the snow at the base of its trunk, the coyote rejoices in the leftovers he will enjoy, the cowherd fears for his finances. The hunter sharpens his weapons. The mountain understands the 'hidden meaning' of the howl. Leopold suggests that the cowherd should 'think like a mountain'.[1]

Who has ever thought like an iceberg?

Let's imagine one last time the young Inuit imitating an iceberg in front of his friends sitting in a circle. In the igloo, seal oil is burning on the soapstone. His mother regularly snips the wick of dried moss so that the smoke does not extinguish it. A soft light spreads, shadows float in the warm atmosphere. The boy stands. He rests his forearms across his chest, lifting his shoulders a little, and begins to bend over. His body moves closer to the ground. He defies gravity. His voice makes slight squeaks. Then a cry comes from his mouth. He falls suddenly. The iceberg has just fallen into the ocean, causing the water to crest as if it were breaking on a reef.

Hiccups, snores and various gurgling sounds suggest that the boy is imitating the raging waves on his back. He restores his balance by banging the ground with his elbows. He then stretches his arms and starts to shear the air with grace. His limbs have become the pectorals of a whale calf. He twists subtly, leaning on his joined feet as if on a caudal. He is now flat on his stomach and shows his back. He opens his mouth. The waves are high. It is not easy to breathe. He finally rolls over and slowly stands up.

His eyes widen and his face turns in all directions. He has become an iceberg again. He is on the lookout because he has heard a noise in the distance. It is a motor. A boat is coming. His gaze fixes on a point above the heads of his assembled friends. He frowns. He is going to get angry. In order to frighten the members of crew approaching him, he manipulates his vocal cords to produce an ominous growl. He rests his cheek on his shoulder. His left side bows. Suddenly, the spectators roar and he throws his hands down in a violent gesture, as if he had to get rid of a burden. A heavy piece of ice breaks away from his sides. Bits of it brush the visitors. Then there is a long silence. The young boy turns and walks away towards the bottom of the igloo, where the shadows reign. The iceberg enters the mist and begins to drift. The theatre is over. Everyone applauds the performance. The artist is congratulated.

We are in the boat. From the deck, once we got over our fright, we interpret the departure of the iceberg as a banishment. Our eyes are clouded with melancholy. Since we do not attribute any concrete intention to them, the blocks seem to us to be emblems of solitude. We look at them in the way we look at the details of a seventeenth-century still life. A basket of old apples, wilted flowers, a burnt-out candle, an hourglass running out. They reflect the image of an inevitable end. The painting by Frederic Edwin Church which opens this book says the same thing. Mirrors of our states of mind, icebergs are allegories of the passing of time, lures in our mundane life. They are beautiful vanities.[2]

We deduce from the fate of the iceberg that the coldness of the ice extinguishes the glow of life. The frosty mists frighten us. They also reassure us. The slight shiver we feel comes from the fact that we do not live in these harsh lands and that we contemplate their dramas from more welcoming shores. We have constructed a scene and prefer the artifice of the show to the realities behind the decor. We can guess that what happens to them affects our lives. But they are out there. We look first deep into the icebergs for evidence of our mortal condition. We then turn our face to one of the surfaces and see only our reflection. Narcissus is no longer a mythological character, as in Ovid's *Metamorphoses*.

Our narcissism condemns us to the cult of appearances.

The illusion is strange. The iceberg sliding across the ocean is a biotic environment in its own right. We understand this with difficulty because life develops below the waterline of the berg. We would have to put on thermal underwear and a wet suit. Diving into the cold waters, we would see it up close. You could walk beside its submerged walls. You could see organisms clinging to the little ice cups carved out by the saltwater currents. The polar cod would open its mouth facing upwards to pick up the suspended plankton. The bearded seal would come up from the depths and grab the cod. Then it would pull itself up onto a ledge to bask in the sunlight, vocalising.

Within seconds, the illusion of a dead block would be reversed.

Any iceberg that breaks off becomes a biological ark in no time. The cracking of its ice is like the howling of a wolf. It reveals the diversity of life forms. The iceberg brings people together around it. Why picture its birth as a sad and lonely fate? Is the ocean so ghostly to us? Let us awaken our pelagic consciousness. Icebergs carried by marine currents are not decorative elements in vanitas paintings, or images of solitude, but essential actors in the primordial cycles of life. The inertia of icy worlds is a misnomer. None of the places we call 'desert' are in fact deserted.

Icebergs illustrate a wild life at work everywhere. They share with animals the same art of appearing and disappearing. As we know, wild life is not always on display. The philosopher Irene J. Klaver reminds us that 'to be wild is to stand out *and* to disappear.'[3] The animal shies away so as not to be devoured. It does not get too close in order to see better. It conceals itself in order to interpret the ways in which other beings inhabit their specific environment. Camouflaging oneself means erasing the contours of one's body, blending into the context, moving without being perceived, and then reappearing at the appropriate moments.

Being wild means knowing how to be discreet.

The iceberg also unites the visible and the invisible. Its identity is not limited to its appearance, nor its value to its visibility. We know how voluminous its underwater part is. Yet it always ends up turning over. What was visible disappears. And what remained hidden shows itself. Each iceberg plays a game of fleeting appearances. Its vital centre depends on the rotating movements that bring the submerged volumes to the surface and engulf the others, so that the visible and invisible sides are never the same. To see an iceberg is thus to see the visible and the invisible in alternation. Nothing is immobile, even if everything seems frozen to the eye that remains on the surface.

Like all living things, the iceberg is characterised by the way it expresses itself, moves, is born and dies. It falls with a crash into the water, only to disappear into anonymity a few months or years later. In the meantime, it shows a part of itself and hides the rest from us. Then it reverses this perceptual pattern by rolling over itself. Like the wolf, the deer, and so many other animals, it is distinguished by its ability to conceal itself. Even when it is huge and very apparent, it knows how to hide from view. It suddenly disappears into the mist, almost silently. It reappears all the more strikingly when it is no longer expected. Its presence is paradoxically light. It slips away and leaves other things to live their own presences.

Basically, icebergs do not 'need' us. Above all, they need us to 'disappear' from time to time.[4] This wild part is imitable. We could learn to develop an art of withdrawal that would be 'an experience in the midst of and with beings and things'. This art would require, for a time, 'laying down all sovereignty in order to open up to the unlimited possibilities of anonymous life.'[5] If we became more discreet, we would in this way be wilder. If we were less visible, we would be more faithful to our principle of freedom. If discretion were our hallmark, we would undoubtedly inhabit environments populated by fabulously varied beings. We would finally imagine 'how much less lonely the world would be'.[6]

Are icebergs too far away for us to be conscious enough of them? It is a reasonable objection. How can we feel close to entities that do not move at the same latitudes as most of the planet's inhabitants? How can we grasp the fact that our personal existences here are linked to their life cycles there? The gap is difficult to bridge. Abstract explanations are not enough. It can only be overcome if we interpret this distance as a figure of continuity; us being part of a common world in which each depends on the other. Icebergs are not solitary. They exist 'by themselves' by being 'well connected' to other beings. They are 'among us', with us, not far away in a wilderness devoid of life.[7]

But there are more and more of them among us. The more they flood the fjords, the more their quantity is a sign of an anomaly. Their increase precedes their disappearance. The more heralds the less. This is the current vertigo, the equation of our near future. The glaciers are tired and the eternal snows are melting. The polar waters are turning blue. The Arctic is gradually becoming a great Mediterranean.

Considering the facts, icebergs are the best teachers. They teach us that every being is a world that brings together other beings and conjugates other worlds. They remind us that life teems in the most seemingly empty places. They invite us to make ourselves indistinguishable at times in order the better to coexist with all living things. Icebergs are discreet colossi, antidotes to narcissism. In them lies the preservation of the world.

It is time to think like an iceberg.

Epilogue

Return to the Ocean

Captain James Cook and his crew are long gone. 'The Impassable' is gone, as are all the icebergs that formed our ice field to the mainland. I am one of her distant granddaughters. The sun's discs have fallen one after the other on the other side of the horizon. Many ships have passed. More sailors came.

They are dressed in all colours and scrutinise me like doctors. They no longer kneel down to tame their fears. They measure my volumes and guess my thoughts with astonishing precision. I feel as if I am ill. They take care of me. They want to know if I have stayed the same since I left my glacier or if I am becoming someone else. My name is strange. A letter, a number: B-49. It is a name without colour or sound. It clatters in the air but does not echo. I am a dot in the square of a map. My trajectory is drawn within an area displayed on an office wall.

I broke away with a crash and a bit of nostalgia. One of my cousins, named A-68, felt the same way. For her, it had all happened three years earlier up north, on the side of the 'Larsen C' ice shelf – according to another captain who had come to visit us more than a century ago and who had wintered, without really wanting to, on a volcanic island. She was closely observed. First a few cracks, then a long crevasse that snaked towards the sea and thunderous roars as she broke apart.

Like so many others before us, the pressure of our ice was too great and the bite of the salt water too painful. We followed the general trend and poured into the ocean like old ladies departing from their tribe to give up the ghost far from kindred regards. I drifted in the Amundsen Sea. A-68, more travelled, went up to the island of South Georgia. Her trajectory frightened those who plotted it with machines whirling overhead. They feared that it would block access to food for penguins, seals and elephant seals if it hit land. It must be said that she was enormous. Next to her, I am so small.

Size does not matter, we have the same destiny. We are getting thinner by the minute, no matter how old our insides are. Snow, water, wind and high temperatures at sea hollow out our forms. We feel the warmth inside us. Our surface ice has melted and refrozen so many times that it has packed our ancient crystals together. Shades of deep cobalt blue are emerging. We see organisms of all kinds pouring in. They surround us, making the water bubble. They swim in the waves that beat against our shores and give off a fibrous light. The currents carry us away. We are heavy, we break, pieces of ice escape from our sides. We are crumbling.

One morning we lose our balance. We roll over and we know that our days are numbered, that there is not much time left to live. We roll over again, then again and again. The distance between our sail and our keel is getting smaller and smaller. But what is happening? Now I'm talking like one of Cook's sailors! And we are not freshwater tanks!

None of us knows any more when we were born, nor from which cluster of snowflakes we came. Like all our parents, we have fallen from the sky and will disappear into the sea. Little pale masses that will never see their people again.

Disaggregated, eclipsed, dissolved, evaporated! Once again we are becoming drops mixed into the undulating waves, foam abandoned to the winds, our nutritive salts stirred up in great columns of water. The voice that speaks to you is no longer mine. It is that of the continent to which we belong. It is that of the ocean to which we return. The continent is our foundation, the land that extends our thoughts. The ocean is our link, the soul that weaves and reweaves our memories. We are the continent and the ocean is us.

If you see our children floating, you are entering a world where you will have to think like us.

We are not landscapes.

We are the past, the present and the future of the world.

Notes

Chapter 1 Through the Looking Glass

1 Elisha Kent Kane, *The US Grinnell Expedition in Search of Sir John Franklin: A Personal Narrative*, New York: Harper & Brothers, 1854; also his *Arctic Explorations: The Second Grinnell Expedition in Search of Sir John Franklin, 1853, '54, '55*, Philadelphia: Childs & Peterson, 1856.

2 Francis Leopold McClintock, *The Voyage of the Fox in the Arctic Seas: A Narrative of the Discovery of the Fate of Sir J. Franklin and his Companions*, London: John Murray, 1859.

3 Louis Legrand Noble, *After Icebergs with a Painter: A Summer Voyage to Labrador and around Newfoundland*, New York: D. Appleton, 1861, p. 47.

4 Ibid., p. 28.

5 Timothy Mitchell, 'Frederic Church's *The Icebergs*: Erratic Boulders and Time's Slow Changes', *Smithsonian Studies in American Art*, 3/4 (1989), pp. 14–17.

6 Cited ibid., pp. 7–8. For theories of the sublime, see Edmund Burke, *A Philosophical Enquiry into the Sublime and Beautiful*, London: Penguin, 1999; Immanuel Kant, *Critique of Judgement*, trans, J. C. Meredith, Oxford: Oxford University Press, 2007,

and *Observations on the Feeling of the Beautiful and Sublime and Other Writings*, Cambridge: Cambridge University Press, 2012.

7 *Navigatio sancti Brendani: alla scoperta dei segreti meravigliosi del mondo*, ed. Giovanni Orlandi and Rossana E. Guglielmetti, Florence: Edizioni del Galluzzo, 2014, chap. XXII, pp. 80–4.

8 Thomas M'Keevor, *A Voyage to Hudson's Bay during the summer of 1812: containing a particular account of the icebergs and other phenomena which present themselves in those regions; also, a description of the Esquimeaux and North American Indians; their manners, customs, dress, language, &c.. &c., &c.* London: Printed for Sir Richard Phillips and Co., 1819, pp. 9–10.

9 Coll Thrush, 'The Iceberg and the Cathedral: Encounter, Entanglement, and Isuma in Inuit London', *Journal of British Studies*, 53/1 (2014), pp. 59–79, at p. 73.

10 Shane McCorristine, *Spectral Arctic: A History of Dreams and Ghosts in Polar Exploration*, London: University College London Press, 2018, p. 9.

11 Christoph Ransmayr, *The Terrors of Ice and Darkness*, trans. J. E. Woods, London: Harper Collins, 1992, p. 22. See also 'Second birthday' in his *Atlas of an Anxious Man*, trans. S. Pare, Chicago: Seagull Books, 2020.

12 Ransmayr, *The Terrors*, p. 52.

13 Ibid., pp. 52, 77, 109.

14 Ibid., p. 25.

15 McCorristine, *Spectral Arctic*, p. 5. Already in 1814, William Scoresby stressed that polar anxiety comes in part from 'the reciprocal action of the ice and the sea' that breaks up whalers (*An Account of the Arctic Regions, with a History and Description of the Northern Whale-Fishery*, Edinburgh: A. Constable and Co., 1820, Vol. I, p. 301). On the

obsession with the poles, see Roland Huntford, *Scott and Amundsen: Their Race to the South Pole*, London: Abacus, 2002; Robert McGhee, *The Last Imaginary Place: A Human History of the Arctic World*, Chicago: University of Chicago Press, 2005; Apsley Cherry-Garrard, *The Worst Journey in the World*, London: Carroll & Graf, [1922] 1989; Pierre Déléage, *La Folie arctique*, Brussels: Zones sensibles, 2017.

16 Arthur Conan Doyle, *Dangerous Work: Diary of an Arctic Adventure*, Chicago: University of Chicago Press, 2012, p. 333.

17 Hans Blumenberg, 'Prospect for a Theory of Nonconceptuality', in *Shipwreck with Spectator: Paradigm for a Metaphor of Existence*, trans. Steven Rendall, Cambridge, MA: MIT Press, 1997, pp. 81–102, at p. 83.

18 Ibid., p. 96.

19 Walter Benjamin, *The Origin of German Tragic Drama*, trans. John Osborne, London: Verso, 1985, p. 166.

20 The term 'Moderns', in this capitalised form, is referring to Bruno Latour's specific 'constitution' of modernity in *We have Never been Modern*, trans. C. Porter, Cambridge, MA: Harvard University press, 1993 [Trans.].

21 Bernard Moitessier, *The Long Way*, trans. W. Rodarmor, New York: Sheridan House, 2003, eBook, p. 214.

22 Ibid., p. 149.

23 Ibid., p. 194.

Chapter 2 The Eye of the Glacier

1 Eugène Rambert, 'Le Voyage du glacier', *Revue des Deux Mondes*, LXXII (1867), pp. 377–410.

2 Martin de La Soudière, *Quartiers d'hiver: ethnologie d'une saison*, Paris: Créaphis, 2016, pp. 39–48.

3 Élisée Reclus, *Histoire d'une montagne*, Paris: Arthaud, [1880] 2017, p. 302.

4 Ibid., pp. 302–3.

5 Ibid., p. 299.

6 Ibid., p. 323.

7 Ibid., p. 321.

8 Mariana Gosnell, *Ice: The Nature, the History and the Uses of an Astonishing Substance*, New York: Alfred A. Knopf, 2005, pp. 94–5.

9 Louis Agassiz, *Études sur les glaciers*, Neuchâtel: Imprimerie O. Petitpierre, 1840, p. 63.

10 Louis Legrand Noble, *After Icebergs with a Painter: A Summer Voyage to Labrador and around Newfoundland*, New York: D. Appleton, 1861, p. 247.

11 Ibid., pp. 248–63.

12 Ibid., pp. 244–5.

13 Ibid., p. 217.

14 John Muir, *Travels in Alaska*, Boston: Houghton Mifflin, 1915, p. 58; emphasis added.

15 Ibid., pp. 67–8.

16 For a brief history of scientific debates, from the eighteenth century to the present day, on how glaciers move, see Frédérique Rémy and Laurent Testut, 'Mais comment s'écoule donc un glacier? Aperçu historique', *CR Geoscience*, 338/5 (2006), pp. 368–85.

17 Noble, *After Icebergs*, p. 272.

18 Muir, *Travels in Alaska*, p. 110; emphasis added.

19 Ibid., pp. 263, 277.

20 Ibid., p. 230; emphasis added.

21 Ibid., p. 284.

22 John Towson, *Icebergs in the Southern Ocean*, Liverpool, privately pubd, 1859, pp. 2–3. The author observes that the size of certain icebergs in the southern hemisphere exceeds 'the limits of credibility' (p. 4).

23 Hinrich Rink, 'On the Large Continental Ice of Greenland, and the Origin of Icebergs in the Arctic Seas', *Journal of the Royal Geographical Society of London*, 23 (1853), pp. 150, 152–3.

24 Henry T. Cheever, *The Whale and his Captors; or, The Whaleman's Adventures, and the Whale's Biography as Gathered on the Homeward Cruise of the 'Commodore Preble'*, New York: Harper & Brothers, 1850, p. 202.

25 N. S. Shaler, 'Icebergs', *Scribner's Magazine*, August 1892, pp. 181–201 (Corpus of Historical American English – COHA). For this and the three previous sources, go to the sites *English Language and Usage Stack Exchange* (english.stackexchange.com/questions /346040/its-easy-to-track- down-the-etymology-of-the-verb-to-calve-what-is-the-origin) and *Word Reference* (forum.wordreference.com/threads/calving-ice-berg -in-danish-and-greenlandic.2624835). NB: The term 'Inuit' is preferred over the exonym 'Eskimo'.

26 In Qaanaaq, in the northwest of Greenland, the word they use today for 'calving' is *uukkartoq*, which refers to a block that breaks away and falls vertically. See Shari Fox Gearheard, Lene Kielsen Holm, Henry Huntington, Joe Mello Leavitt, Andrew R. Mahoney, Margaret Opie, Toku Oshima and Joelie Sanguya, eds, *The Meaning of Ice: People and Sea Ice in Three Arctic Communities*, Totnes: International Polar Institute Press, 2013, p. 336.

27 Jane Bennett's phrase in *Vibrant Matter: A Political Ecology of Things*, Durham, NC: Duke University Press, 2010.

28 Here I am following the works of Stephen J. Pyne, *The Ice*, London: Weidenfeld & Nicolson, 2003, pp. 7–20 and 125–8, and Douglas I. Benn and Jan A. Åström, 'Calving Glaciers and Ice Shelves', *Advances in Physics: X*, 3/1 (2018), pp. 1048–76. For further details, see Douglas I. Benn and David J. Evans, *Glaciers and*

Glaciation, London: Hodder Education, 2010, and Grant R. Bigg, *Icebergs: Their Science and Links to Global Change*, Cambridge: Cambridge University Press, 2016.

29 Some circumstances produce chain reactions of icebergs that rotate and disintegrate entire shelves; see Douglas MacAyeal, Ted Scambos, Christina Hulbe and Mark Fahnestock, 'Catastrophic Ice-Shelf Break-Up by an Ice-Shelf-Fragment-Capsize Mechanism', *Journal of Glaciology*, 49/164 (2003), pp. 22–36.

30 In Antarctica, the letters are distributed geographically: 'A' designates the zone extending from the Bellingshausen Sea to the Weddell Sea, 'B' goes from the Amundsen Sea to the east of the Ross Sea, 'C' from the west of the Ross Sea to the Wilkes Sea, 'D' from the Amery Barrier to the east of the Weddell Sea. The numbers indicate the order of appearance of the icebergs. The names are assigned by the US National Ice Center (NIC).

31 From a linguistic point of view, a metaphor is an implied, or abbreviated, comparison because it does not use comparative words.

32 Werner Herzog, dir., *Encounters at the End of the World*, 2007.

33 Val Plumwood, *The Eye of the Crocodile*, Canberra: Australian National University Press, 2012; DOI: http://doi.org/10.22459/EC.11.2012.

Chapter 3 Unexpected Lives

1 Barry Lopez, *Arctic Dreams: Imagination and Desire in a Northern Landscape*, London: Harvill Press, 1999.

2 Cited in Stephen J. Pyne, *The Ice*, London: Weidenfeld & Nicolson, 2003, p. 117.

3 Christoph Ransmayr, *The Terrors of Ice and Darkness*,

trans. J. E. Woods, London: Harper Collins, 1992, pp. 72–3.

4 Charles Weinstein, *Arctique extrême: les Tchouktches du détroit de Béring*, Paris: Autrement, 1999, p. 144.

5 See Anon, *The King's Mirror (Speculum regale-Konungs skuggsjá)*, trans. Laurence Marcellus Larson, www.gutenberg.org/files/61264/61264-h/612 64-h.htm.

6 Lopez, *Arctic Dreams*, pp. 123, 213–14.

7 Cited in Pyne, *The Ice*, pp. 67–8.

8 Barry Lopez, *Horizon*, London: Alfred A. Knopf, 2019, pp. 447–52. On the forms of life that depend on icebergs, see also Kenneth L. Smith, Alana D. Sherman, Timothy J. Shaw and Janet Sprintall, 'Icebergs as Unique Lagrangian Ecosystems in Polar Seas', *Annual Review of Marine Science*, 5/1 (2011), pp. 269–87.

9 Lopez, *Arctic Dreams*, pp. 252–3.

10 Gretel Ehrlich, *In the Empire of Ice: Encounters in a Changing Landscape*, Washington, DC: National Geographic Society, 2010, p. 270.

11 In the end, Lopez's ultimate desire is 'to be with icebergs'; see *Arctic Dreams*, p. 205.

12 Camille Seaman's photographs are published in *Melting Away: A Ten-Year Journey through Our Endangered Polar Region*, Princeton, NJ: Princeton Architectural Press, 2014. The video of her TED talk of 16 June 2011 can be seen at: www.ted.com /talks/camille_seaman_haunting_photos_of_polar_ice.

13 Cited by Romain Bertrand, *Le Détail du monde: l'art perdu de la description de la nature*, Paris: Seuil, 2019, p. 183.

14 Robin Wall Kimmerer, 'Learning the Grammar of Animacy', in *Braiding Sweetgrass: Indigenous Wisdom, Scientific Knowledge, and the Teachings of Plants*, Minneapolis: Milkweed Editions, 2013, p. 55; original emphasis. In linguistics, nouns are 'animate' if their

referent is considered alive. Kimmerer extends the utility and meaning of this category by making it a distinctive feature of indigenous cosmology, a reasoning we are following here.

15 Ibid., p. 55.

16 Ibid., p. 57. The French language uses personal pronouns for things and people while separating the roles. Speaking of an apple tree, one would say that 'he' is flowering, but one would more often think that 'it' is a tree. English, for its part, uses 'it' in both cases.

17 David Abram, *The Spell of the Sensuous: Perception and Language in a More-than-Human World*, New York: Random House, 1996, p. 157.

18 Lopez, *Arctic Dreams*, p. 29.

19 It is Barry Lopez quoting Franz Boas in *Arctic Dreams*, p. 265 (and p. 285 for Kane's comparison of the sounds of puppies and bees).

20 Ibid., p. 265.

21 Ibid., pp. 290–1.

22 Edmund Carpenter, *Eskimo Realities*, New York: Rinehart & Winston, 1973, p. 21 (the author is relating his field experiences in the 1950s).

23 Barry Lopez, *Horizon*, London: Alfred A. Knopf, 2019, pp. 167–70.

24 Philippe Descola calls this cognitive process *mondiation* (worlding); he classifies this into four principle forms: animism, naturalism, totemism and analogism. See 'Cognition, Perception and Worlding', *Interdisciplinary Science Reviews*, 35/3–4 (2010), pp. 334–40; and 'The Ways of the World', in *Beyond Nature and Culture*, trans. J. Lloyd, Chicago: University of Chicago Press, 2005, pp. 247–89.

25 Tim Ingold, 'Rethinking the Animate, Re-animating Thought', *Ethnos*, 71/1 (2006), pp. 10 and 14.

26 Carpenter, *Eskimo Realities*, pp. 26–7.

27 John Muir, *Travels in Alaska*, Boston: Houghton Mifflin, 1915, pp. 229 and 230.

28 'Soundscape' is an expression coined by the composer and teacher Raymond Murray Schafer in *The Soundscape: Our Sonic Environment and the Tuning of the World*, New York: Alfred A. Knopf, 1977. Since then there has been a new thread of work in the 'ecology of sound landscapes'. See, for example, Bryan C. Pijanowski, Almo Farina, Stuart H. Gage, Sarah L. Dumyahn and Bernie L. Krause, 'What is Soundscape Ecology? An Introduction and Overview of an Emerging New Science', *Landscape Ecology*, 26/9 (2011), pp. 1213–32.

29 It was James Balog who came up with the concept of 'visual voice'. With his fixed cameras, he filmed the slow-motion disintegration of glacial ecosystems under the influence of climate deregulation. See his documentary *Chasing Ice*, 2012.

30 Such are the conclusions drawn in the articles by Robert P. Dziak, Matthew J. Fowler, Haruyoshi Matsumoto, DelWayne R. Bohnenstiehl, Minkyu Park, Kyle Warren and Wong Sang Lee, 'Life and Death Sounds of Iceberg A53a', *Oceanography*, 26/2 (2013), pp. 10–12; Haruyoshi Matsumoto, DelWayne R. Bohnenstiehl, Jean Tournadre, Robert P. Dziak, Joseph H. Haxel, T.-K. A. Lau, Matthew J. Fowler and Sigrid A. Salo, 'Antarctic Icebergs: A Significant Natural Ocean Sound Source in the Southern Hemisphere', *Geochemistry, Geophysics, Geosystems*, 15/8 (2014), pp. 3448–58; Läslo G. Evers, David N. Green, N. W. Young and Mirjam Snellen, 'Remote Hydroacoustic Sensing of Large Icebergs in the Southern Indian Ocean: Implications for Iceberg Monitoring', *Geophysical Research Letters*, 40 (2013), pp. 4694–9. Spectrograms can also be consulted on the Pacific Marine Environmental Laboratory site: www.pmel.noaa.gov/acoustics/sounds_cryogenic.html.

31 See Philippe Blondel's site: www.bath.ac.uk/case-studies/listening-to-icebergs-on-climate-change.

32 The name 'white vagabond' seems to echo Moby Dick who erupted, in the final chapter of Herman Melville's novel, spreading the waves before him like an iceberg 'swiftly rising to the surface'. *Moby Dick, or, The Whale*, ed. Mary K. Bercaw Edwards, New York: Penguin, 2009, Chap. 135, 'The Chase', p. 902.

33 Carpenter, *Eskimo Realities*, pp. 33–6.

34 Lopez, *Arctic Dreams*, p. 276.

35 Maurice Merleau-Ponty, *L'Oeil et l'esprit*, Paris: Gallimard, 1964, p. 85.

Chapter 4 Social Snow

1 Bernard Saladin d'Anglure, *Inuit Stories of Being and Rebirth: Gender, Shamanism, and the Third Sex*, trans. Peter Frost, Winnipeg: University of Manitoba Press, 2018, p. 50. eBook. Rite recorded in the spring of 2004.

2 Paul-Émile Victor and Joëlle Robert-Lamblin, *La Civilisation du phoque: légendes, rites et croyances des Eskimo d'Ammassalik*, Bayonne: Raymond Chabaud, 1993, p. 320. The first spell was reported by an informer called Abudu. The second was recorded at Sermiligak in March 1935.

3 Ibid., p. 199 ('Poèmes d'Ammassalik').

4 Saladin d'Anglure, *Inuit Stories*, p. 12. This account was transcribed by Eber and Pitseolak in 1975.

5 Eduardo Viveiros de Castro, *Cannibal Metaphysics*, trans. Peter Skafish, Minneapolis: Univocal, 2014; and 'Cosmological Deixis and Amerindian Perspectivism', *Journal of the Royal Anthropological Institute*, 4/3 (1998), pp. 469–88.

6 Saladin d'Anglure, *Inuit Stories*, p. 16.

7 Ibid., p. 17.

8 This redefines the play of kinship, as Jean Malaurie reminds us: 'The name relates and aligns. A child bearing the name of his grandfather will not be addressed

as "my son" by his father, but "my father", even if he is very young. A girl given the name of her grandfather will be called "my father" by her father.' *Les Derniers Rois de Thulé*, Paris: Plon, 1989, p. 194.

9 Bérengère Cournut, *De pierre et d'os*, Paris: Le Tripode, 2019, pp. 47–8.

10 Ibid., p. 49.

11 Saladin d'Anglure, *Inuit Stories*, pp. 40–1.

12 Ibid., p. 55.

13 Mario Rigoni Stern, 'Neige', in *Sentiers sous la neige*, trans. Monique Baccelli, Lyon: La Fosse aux ours, 2000, pp. 89–93. On words used to 'speak of snow', see also Martin de La Soudière, *Quartiers d'hiver: ethnologie d'une saison*, Paris: Créaphis, 2016, pp. 49–62.

14 These examples come from the following three works: Shari Fox Gearheard, Lene Kielsen Holm, Henry Huntington, Joe Mello Leavitt, Andrew R. Mahoney, Margaret Opie, Toku Oshima and Joelie Sanguya, eds, *The Meaning of Ice: People and Sea Ice in Three Arctic Communities*, Totnes: International Polar Institute Press, 2013, pp. 132–3 ('Terminology, Characteristics, and Change of the Seasons at Kangiqtugaapik') and pp. 322–37 ('Sea Ice Terminology'); Igor Krupnik, Claudia Aporta, Shari Gearheard, Gita J. Laidler and Lene Kielsen Holm, eds, *SIKU: Knowing Our Ice: Documenting Inuit Sea Ice Knowledge and Use*, Berlin: Springer, 2010, Appendix A, pp. 453–63; Louis-Jacques Dorais, 'Les Mots inuits pour la neige et la glace', www.thecanadianencyclopedia.ca/fr/article/les-mots-inuits-pour-la-neige-et-la-glace. It goes without saying that the translations given here of the meanings of these terms comes from the translations in the texts cited. I have left out the inverted commas to assist legibility.

15 Shari Gearheard, 'What's in a Word? The High-Stakes Ties of Language, Knowledge and Environment', https://bifrostonline.org/whats-in-a-word-the-high-stakes-ties-of-language-knowledge-and-environment/.

16 Edmund Carpenter, *Eskimo Realities*, New York: Rinehart & Winston, 1973, p. 43.

17 Barry Lopez, *Arctic Dreams: Imagination and Desire in a Northern Landscape*, London: Harvill Press, 1999, p. 276. Marcel Mauss wrote in the years 1904–5 about the Inuit: 'Each season serves to define an entire class of beings and objects.' *Seasonal Variations of the Eskimo: A Study in Social Morphology*, trans J. J. Fox, London: Routledge, 1979, pp. 61–2.

18 Lopez, *Arctic Dreams*, p. 7.

19 Jean Malaurie's expression, in *Les Derniers Rois*, p. 159.

20 For the most part, ice spirits are malevolent or very worrying. Such is the case for the *witiko* in the 'ice dreams' of the Cree Indians of James Bay (Robert A. Brightman, *Grateful Prey: Rock Cree Human–Animal Relationship*, Los Angeles: University of California Press, 1993, pp. 152–6). Or for the 'ice bear' among the Gwich'in of Alaska (Nastassja Martin, *Les Âmes sauvages: face à l'Occident, la résistance d'un peuple d'Alaska*, Paris: La Découverte, 2016, pp. 196–201).

21 Victor and Robert-Lamblin, *La Civilisation du phoque*, p. 299 (the spell was provided by Widimi from Kulusuk).

22 Henry P. Huntington, Shari Gearheard, Lene Kielsen Holm, George Noongwook, Margaret Opie and Joelie Sanguya, 'Sea Ice is Our Beautiful Garden: Indigenous Perspectives on Sea Ice in the Arctic', in David N. Thomas, ed., *Sea Ice*, Chichester: John Wiley & Sons, 2017, p. 590 and pp. 586–7.

23 Vincent Descombes provide this definition of 'form of life' in *Les Institutions du sens*, Paris: Minuit, 1996, p. 93.

24 Gearheard et al., *The Meaning of Ice*, p. 63.

25 Ibid., pp. 135 and 137. In the far North, winter is eagerly awaited. The mud of other seasons is cause for concern. Everywhere, it is the frozen land that makes

travel possible. In Canada, too, '"snowy weather" is considered a time for travel ... Winter is not the reclusive season we imagine it to be; in the past, people travelled more quickly than in summer.' Pierre Deffontaines, *L'Homme et l'hiver au Canada*, Paris: Gallimard, 1957, p. 155.

26 Gearheard et al., *The Meaning of Ice*, p. 137.

27 Guy Mary-Rousselière, *Qitdlarssuaq: l'histoire d'une migration polaire*, Paris: Paulsen, 2008, p. 44. The rest of the story says that Pinaitsoq jumps from ice block to ice block, feeds on the remains of seals and manages to reach a new piece of ice that brings him back to land. This book recounts the adventure begun in 1856 by the shaman Qitdlarssuaq, who wanted to find out if other Inuit lived to the north. With about forty companions, he left Baffin Island. They met some near Siorapaluk and lived with them for almost six years.

28 Ibid., p. 107.

29 Paul-Émile Victor, *Boréal: une année en pays esquimau*, Paris: Points, [1936] 2013, pp. 110–11.

30 Paul-Émile Victor, *Banquise*, Paris: Grasset, 1939, p. 214.

31 Charles Ferdinand Ramuz, *La Grande Peur dans la montagne*, Paris: Le Livre de poche, 2011, p. 78 and pp. 170–7.

32 Julie Cruikshank, 'Glaciers and Climate Change: Perspectives from Oral Tradition', *Arctic*, 54/4 (2001), pp. 384–5 and 387–8. See also, by the same author, *Do Glaciers Listen? Local Knowledge, Colonial Encounters and Social Imagination*, Vancouver: University of British Columbia Press, 2005, p. 39.

33 Cruikshank, *Do Glaciers Listen?*, pp. 73–4.

34 Ann Fienup-Riordan and Eddy Carmack, '"The Ocean is Always Changing": Nearshore and Farshore Perspectives on Arctic Coastal Seas', *Oceanography*, 24/3 (2011), pp. 266–79, at p. 267. The interview took place in March 2007.

35 Ibid., pp. 269–70.

36 Ibid., pp. 275–6 (original emphasis).

37 Ann Fienup-Riordan, 'Yup'ik Perspectives on Climate Change: "The World is Following its People"', *Inuit Studies*, 34/1 (2010), p. 58.

38 Geremia Cometti, *Lorsque le brouillard a cessé de nous écouter: changement climatique et migrations chez les Q'eros des Andes péruviennes*, Berne: Peter Lang, 2015, pp. 152–8.

39 Ibid., pp. 209–11.

40 Claude Lévi-Strauss, 'The Canoe Journey of the Sun and the Moon', in *Introduction to a Science of Mythology*, vol. 3: *The Origin of Table Manners*, trans. John and Doreen Weightman, London: Jonathan Cape, 1978, pp. 137–82.

41 John Muir, *Travels in Alaska*, Boston: Houghton Mifflin, 1915, p. 146.

42 Julie Cruikshank, 'Are Glaciers "Good to Think With"? Recognising Indigenous Environmental Knowledge', *Anthropological Forum*, 22/3 (2012), pp. 242–3.

43 Joelie Sanguya's expressions in Gearheard et al., *The Meaning of Ice*, pp. 135 and 137.

44 Victor and Robert-Lamblin, *La Civilisation du phoque*, pp. 392–3.

45 Cruikshank, *Do Glaciers Listen?*, pp. 114ff. (and her article 'Legend and Landscape: Convergence of Oral and Scientific Traditions in the Yukon Territory', *Arctic Anthropology*, 18/2 (1981), p. 76).

46 David Abram, *Becoming Animal: An Earthly Cosmology*, New York: Random House, 2010, pp. 173–4.

Chapter 5 A Less Lonely World

1 On different techniques of dating, see Jean Jouzel, Claude Lorius and Dominique Raynaud, *Planète*

blanche: les glaces, le climat et l'environnement, Paris: Odile Jacob, 2008, pp. 63–166.

2 Gretel Ehrlich, *In the Empire of Ice: Encounters in a Changing Landscape*, Washington, DC: National Geographic Society, 2010, pp. 270–1.

3 Mark Carey, 'History of Ice: How Glaciers Became an Endangered Species', *Environmental History*, 12/3 (2007), pp. 497–527.

4 Mark Carey relates the five criteria to the case of the glaciers covering Mount Kilimanjaro in Tanzania: ibid., p. 513.

5 See the UNESCO site: https://whc.unesco.org/en/ne ws/65.

6 Mark Carey, 'The Trouble with Climate Change and National Parks', *Forest History Today*, 23/1 (2017), pp. 57–67; Julie Brugger, Katherine W. Dunbar, Christine Jurt and Ben Orlove, 'Climates of Anxiety: Comparing Experience of Glacier Retreat across Three Mountain Regions', *Emotion, Space and Society*, 6 (2013), pp. 4–13. M Jackson (Jerilynn "M" Jackson's pen-name) also stresses the 'last chance' aspect of glacier tourism in Iceland: see *The Secret Lives of Glaciers*, Brattleboro, VT: Green Writers Press, 2019, pp. 204ff. The 'neoliberal glacier' expression can be found in Mark Carey's book *In the Shadow of Melting Glaciers: Climate Change and Andean Society*, Oxford: Oxford University Press, 2010 (Chap. VII).

7 See Susan Kollin, 'The Wild, Wild North: Nature Writing, Nationalist Ecologies, and Alaska', *American Literary History*, 12/1–2 (2000), pp. 41–78, at p. 48.

8 Wilderness Act (Public Law 88-577 [16 US. C. 1131-1136], 88th Congress, Second Session, 3 September 1964, p. 1.

9 William Cronon, 'The Trouble with Wilderness: Or, Getting Back to the Wrong Nature', *Environmental History*, 1/1 (1996), pp. 7–28, at p. 7.

10 Irene J. Klaver, 'Wild: Rhythm of the Appearing

and Disappearing', in Michael P. Nelson and J. Baird Callicott, *The Wilderness Debate Rages On: Continuing the Great New Wilderness Debate*, Athens: University of Georgia Press, 2008, p. 486.

11 Cronon, 'The Trouble with Wilderness', p. 13, pp. 20–21 and p. 17.

12 Julie Cruikshank, *Do Glaciers Listen? Local Knowledge, Colonial Encounters and Social Imagination*, Vancouver: University of British Columbia Press, 2005, p. 255.

13 The quote from Henry D. Thoreau is found in his essay 'Walking', *Atlantic Monthly*, IX/LVI (1862), p. 665. I share Jack Turner's diagnosis of the all too frequent conflation of wilderness and wildness: see 'In Wildness is the Preservation of the World', in J. Baird Callicott and Michael P. Nelson, eds, *The Great New Wilderness Debate*, Athens: University of Georgia Press, 1998, pp. 617–27.

14 For an exhaustive analysis, read the first part of Catherine Larrère and Raphaël Larrère's *Penser et agir avec la nature: une enquête philosophique*, Paris: La Découverte, 2015, pp. 21–151.

15 Ibid., p. 43.

16 Shelley Wright, *Our Ice is Vanishing/Sikuvut Nunguliqtuq: A History of Inuit, Newcomers and Climate Change*, Montreal: McGill-Queen's University Press, 2014, p. 13.

17 On the event and location, see www.notokmo vie.com, as well as articles by Lacy M. Johnson ('How to Mourn a Glacier: In Iceland, a Memorial Ceremony Suggests New Ways to Think About Climate Change', *New Yorker*, 20 October 2019, www.newyorker.com/news/dispatch/how-to-mourn -a-glacier); and by Cymene Howe ('On Cryohuman Relations', in Rafico Ruiz, Paula Schönach and Rob Shields, eds, *After Ice: Cold Humanities for a Warming Planet*, Durham, NC: Duke University Press, 2022).

18 Cited by Jackson, *The Secret Lives of Glaciers*, p. 157.
    Albedo is the 'fraction of solar energy reflected back
    into space.' The more reflective a surface is, the higher
    the albedo. That is, from 80 to 90 per cent in the case
    of fresh snow, between 75 and 80 per cent for sea ice);
    see Christophe Migeon, *Petit manuel du voyageur
    polaire*, Paris: Paulsen, 2014, pp. 54–5.

19 Jackson, *The Secret Lives of Glaciers*, pp. 158ff.

20 Interview with Dögg, ibid., p. 173. When glaciers 'turn
    turquoise' in Iceland, it is said that 'you can expect rain
    or snow' (p. 76).

21 Philippe Descola, *Beyond Nature and Culture*, trans.
    J. Lloyd, Chicago: University of Chicago Press, 2005,
    p. 250.

22 Howe, 'On Cryohuman Relations'. She considers
    that the totality of 'cryohuman relations' should be
    rethought as soon as possible.

23 Kelton Minor, Gustav Agneman, Navarana Davidsen,
    Nadine Kleemann, Ulunnguaq Markussen, Allan Olsen,
    David Dreyer Lassen and Minik Thorleif Rosing,
    *Greenlandic Perspectives on Climate Change 2018–
    2019: Results from a National Survey*, University of
    Greenland and University of Copenhagen, Kraks Fond
    Institute for Urban Research, 2019. The six regions
    studied are: Avannaata, Qeqertalik, Qeqqata, West
    Sermersooq, Kujalleq and East Sermersooq.

24 Dan McDougall, 'Life on Thin Ice: Mental Health at
    the Heart of the Climate Crisis', *The Guardian*, 12
    August 2019.

25 Glenn Albrecht, *Earth Emotions: New Words for a
    New World*, Ithaca, NY: Cornell University Press,
    2019, pp. 38–39.

26 Ibid., p. 42. See also my article 'On n'achève pas un
    glacier qui sauve un peuple', *Socialter*, 7 July 2021,
    www.socialter.fr/article/on-n-acheve-pas-un-glacier
    -qui-sauve-un-peuple.

27 Barry Lopez, *Arctic Dreams: Imagination and Desire*

in a Northern Landscape, London: Harvill Press, 1999, p. 279.

28 An allusion to Davis Guggenheim's 2006 film of the same name, *An Inconvenient Truth*, which draws heavily on Al Gore's speeches and activism.

29 Donna Haraway, 'Anthropocene, Capitalocene, Plantationocene, Chthulucene: Making Kin', *Environmental Humanities*, 6/1 (2015), pp. 159–65. This reference was also used by Howe, 'On Cryohuman Relations'. It seems appropriate to me in the context of today's glacial melting.

30 I borrow this citation from Philippe Descola: 'De la Nature universelle aux natures singulières: quelles leçons pour l'analyse des cultures?', in Descola, ed., *Les Natures en question*, Paris: Odile Jacob and Collège de France, 2018, p. 136. For a discussion of 'collectives', see also, by the same author, *Beyond Nature and Culture*, pp. 247–68.

31 Jackson, *The Secret Lives of Glaciers*, p. 140.

32 Tim Ingold, 'Rethinking the Animate, Re-Animating Thought', *Ethnos*, 71/1 (2006), p. 19.

33 Philippe Descola's expression, in a discussion on glaciers among other entities: *La Composition des mondes: entretien avec Pierre Charbonnier*, Paris: Flammarion, 2017, p. 281.

34 Geremia Cometti, *Lorsque le brouillard a cessé de nous écouter: changement climatique et migrations chez les Q'eros des Andes péruviennes*, Berne: Peter Lang, 2015, p. 229.

35 Larrère and Larrère, *Penser et agir avec la nature*, p. 23.

36 Cruikshank, *Do Glaciers Listen?*, pp. 257, 49 and 64.

37 Karine Gagné, Mattias Borg Rasmussen and Ben Orlove, 'Glaciers and Society: Attributions, Perceptions, and Valuations', *WIRES Climate Change*, 5/6 (2014), pp. 793–808, at p. 804. See also Lill Rastad Bjørst, 'The Tip of the Iceberg: Ice as a Non-Human Actor

in the Climate Change Debate', *Études/Inuit/Studies*, 34/1 (2010), pp. 133–50.

38 See Edward O. Wilson, *Half-Earth: Our Planet's Fight for Life*, New York: Liveright, 2016. Wilson has a 'radical' proposition: reserve half of the planet for biodiversity.

39 Aldo Leopold, *A Sand County Almanac*, Oxford: Oxford University Press, 1949, pp. viii, 202, 204 and 214.

Chapter 6 Thinking Like an Iceberg

1 Aldo Leopold, *A Sand County Almanac*, Oxford: Oxford University Press, 1949, pp. 133, 132.

2 For Henri Michaux, icebergs are 'solitary without need'; see 'Icebergs', in *La Nuit remue*, Paris: Gallimard, 1935. The gloomy atmosphere of the poem echoes Charles Leconte de Lisle's 'Polar Landscape': 'A dead world, immense foam of the sea, / abyss of sterile shadows and spectral gleams, / spurts of convulsive peaks stretched out in spirals / that go madly into the bitter fog' (*Poèmes barbares*, Paris: Gallimard, 1985).

3 Irene J. Klaver, 'Wild Rhythm of the Appearing and Disappearing', in Michael P. Nelson and J. Baird Callicott, eds, *The Wilderness Debate Rages On: Continuing the Great New Wilderness Debate*, Athens: University of Georgia Press, 2008, p. 485; original emphasis.

4 I am transposing Irene Klaver's words to the case of icebergs. Ibid., p. 496.

5 Pierre Zaoui, *La Discrétion, ou L'Art de disparaître*, Paris: Autrement, 2013, p. 135.

6 Robin Wall Kimmerer, 'Learning the Grammar of Animacy', in *Braiding Sweetgrass: Indigenous Wisdom, Scientific Knowledge, and the Teachings of Plants*, Minneapolis: Milkweed Editions, 2013, p. 58.

7 These expressions are from Baptiste Morizot, 'Le Devenir du sauvage à l'Anthropocène', in Rémi Beau and Catherine Larrère, eds, *Penser l'Anthropocène*, Paris: Presses de Sciences Po, 2018, pp. 249 and 257–9. See also my article 'Un iceberg dans la forêt', *Libération*, 2 June 2020, www.liberation.fr/france/2020/06/02/un-iceberg-dans-la-foret_1789543.